INTEGRATING MULTIPLE SOURCES OF INFORMATION FOR IMPROVING HYDROLOGICAL MODELLING: AN ENSEMBLE APPROACH

Isnaeni Murdi Hartanto

INTEGRATING MULTIPLE SOURCES OF INFORMATION FOR IMPROVING HYDROLOGICAL MODELLING: AN ENSEMBLE APPROACH

DISSERTATION

Submitted in fulfilment of the requirements of
the Board for Doctorates of Delft University of Technology
and of
the Academic Board of IHE Delft Institute for Water Education
for the Degree of DOCTOR
to be defended in public
on Tuesday, 19 March 2019 at 12.30 hours
in Delft, the Netherlands

by

Isnaeni Murdi HARTANTO

Master of Science Water Science and Engineering
UNESCO-IHE Institute for Water Education

born in Kendal, Central Java, Indonesia

This dissertation has been approved by the

Promotor: Prof. dr. D.P. Solomatine

Copromotor: Dr. S.J. van Andel

Composition of Doctoral Awarding Committee:

Rector Magnificus TU Delft	Chairman
Rector IHE Delft	Vice-Chairman
Prof. dr. D.P. Solomatine	IHE Delft/TU Delft, promotor
Dr. S.J. van Andel	IHE Delft, copromotor

Independent members:

Prof. dr. ir. N.C. van de Giesen	TU Delft
Prof. dr. ir. A.H. Weerts	Wageningen University and Research
Dr. R.J.J. Neves	Technical University of Lisbon, Portugal
Ing. R. van der Zwan	Principle Water Board of Rijnland
Prof. dr. M. J. Franca	IHE Delft/TU Delft, reserve member

CRC Press/Balkema is an imprint of the Taylor & Francis Group, and informa business

Published by:
CRC Press/Balkema
PO Box 11320, 2301 EH Leiden, the Netherlands
Email: pub.nl@taylorandfrancis.com
www.crcpress.com – www.taylorandfrancis.com
ISBN 978-0-367-26543-4

Acknowledgments

I would like to express my gratitude to those who helped me in finishing my PhD research, without them, this book would not be finished as it is now. These include those who were helping me directly on my research, and those who contributed indirectly through supporting my daily life and keep my moral up.

First of all I like to thank God Almighty who gave me this beautiful experience in doing this research and finishing it.

Great thanks to the MyWater FP-7 EU Project for providing financial support for this research, Rijnland Waterboard for providing data about the Rijnland water system, and SARA Foundation for giving me their permission in using High Performance Computing and Data Infrastructure via SURF-SARA, to run the highly demanding models that are used in this research.

Next is my high gratitude for my promoter Dimitri Solomatine for his willingness to be my supervisor during the research, also my co-promoter Schalk Jan van Andel for his many supports and patiently guiding me throughout this research, also for the ideas, small talks, and many other constructive assists. I wish also to thank: all IHE Delft lecturers, especially Ioana Popescu, Shreedhar Maskey, Andreja Jonoski, Gerald Corzo Perez, Arnold Lobbrecht and Hans van der Kwast in helping me with ideas and suggestions for the research, Ab Veldhuizen and Arjon Buijert who taught me how to understand SIMGRO software, and fellow MyWater project contributors, Thomas, Pedro, Waldenio, Inez, and Diogo with whom I shared knowledge about satellites, hydrological information and models, and IHE staff Jolanda, Marielle, Anique, Sylvia, Jos, and Gerda who always helped me with administrative support during my stay in Delft.

To my fellow Hydroinformatics researchers, Adrian, Anuar, Mario, Pan, Miguel, Juan Carlos, Micah, Yang, Tracy, Maurizio, and Kun, I am thankful for their ideas, scripts, discussions, knowledge sharing and aids during the tiring research and fun little chats during stressful times, in the middle of data processing, Matlab, Python, Ubuntu, remote access, graphs, maps, models, and awkward silent times. Also many thanks to my fellow UN-IHE researchers Shahrizal, Leo, Fiona, Yuli, Linh, Tarn, Clara, Yos, Gladys, Sony, and Mr. Suryadi for accompanying me, in lunch, nonsense talks, no purpose walks, Europe trips, and un-harmful bullying, inside and outside research time.

I am also thankful to the Indonesian researchers from TU Delft, Senot, Pungky, Yazdi, Diecky, Jun, Ida, Adhi, Budi, Marwan, Topan, and their families, and also to other PPI Delft friends that are not mentioned here, for the moments that were shared together in family gatherings and various occasions in The Netherlands, and for the various foods that we shared to soften our crave for Indonesian cooking.

My co-workers and supervisors in the Ministry of Public works, Mr. Hari Suprayogi, Mr. Iwan Nursyirwan, Mr. Mudjiadi, Mr. Agus Suprapto, Mr. Widiarto, Mr. Trisasongko Widianto, Mr. Edy Juharsyah, Mr. and Mrs. Agni Handoyoputro, Mrs. Lilik Retno, I thank for their permission to escape from my duties to do this research. I also send many thanks to my best co-worker and friend Niken for all her support during my Master and PhD.

Lastly, I would like to thank my late Father for his teaching, my caring Mother for her patience, my beloved sisters Evy, Sari and Dian for their pray and in accompanying our mother, my patient wife Lulu for her company and everlasting support during good and bad times, and my children Arham, Alfiqh, Akif, and Althaf who keep my heart and mind up and able to always move forward.

There are definitely some I have missed to mention in this acknowledgement, but your contributions are greatly appreciated.

Jakarta, Indonesia

Isnaeni Murdi Hartanto

Summary

The availability of Earth observation (EO) and numerical weather prediction data for hydrological modelling and water management has increased significantly, creating a situation that today, for the same variable, estimates may be available from two or more sources of information. Precipitation data, for example, can be obtained from rain gauges, weather radar, satellites, or outputs from numerical weather models. Land use data can be obtained from land survey, satellite imagery, or a combination of the two. Each of these data sources provides an estimate of a catchment characteristic and related hydrological model parameters, or of a hydrometeorological variable. Estimates from each data source vary in magnitude or temporal and spatial variability. It is not always possible to judge which data source is the most accurate. One data source may perform poorly in one situation but give an accurate estimate for another. Yet, in hydrological modelling, usually, a particular set of catchment characteristics and input data is selected, possibly ignoring other relevant data sources. One of the reasons may be that despite vast research and development efforts in integration methods for sub-sets of the available data sources, there is no comprehensive data-model integration framework assuming existence and enabling effective use of multiple data sources in hydrological modelling.

The main objective of this thesis, therefore, is to develop such a data-model integration framework, and test it on a case study.

The framework developed, is based on the ensemble prediction approach. An ensemble prediction, as a particular class of probabilistic predictions, gives several forecasts (members) for the same time and location, instead of a single, deterministic, forecast. Multiple predictions are generated to account for the uncertainty in initial conditions, parameter values, forcing, or model structure. When a predicted probability of occurrence is attached to each of the forecasts, a probability distribution can be derived. Ensemble methods have been mostly developed and used in numerical weather prediction. One way of creating an ensemble is by perturbing initial conditions of the meteorological model, and re-running it, each time creating an additional ensemble member. This type of ensemble is called an Ensemble Prediction System (EPS). In another ensemble prediction method, called "poor man's" or multi-model ensemble, the ensemble is generated by selecting several available predictions from different meteorological or hydrological models and providers. The third method for ensemble simulation, is random sampling from the parameter and input space and re-running the model. So far, the third method has been the most widely used method in hydrological ensemble modelling.

The developed framework for incorporating multiple data sources in hydrological modelling uses a method of ensemble prediction that is similar to the multi-model approach. In this framework, each available data source is used to derive catchment parameter values or input time series. Each unique combination of catchment and input data sources thus leads to a different hydrological simulation result: a new ensemble member. Together, the members form an ensemble of hydrological simulations. By following this approach, all available data sources are used effectively and their information is preserved, resulting in a hydrological ensemble simulation that quantifies a certain part of the data and parameter uncertainty. Assigning weights to the ensemble members allows for assessing the probability distribution of the simulation results and its moments, including the ensemble mean.

Next to the effective use of multiple data sources, the framework accommodates for applying multiple data-model integration methods, e.g. data-infilling, data-correction, data-merging, data assimilation, and input-data updating based on model results (feedback loop). Each alternative method of data-model integration leads to yet another unique hydrological simulation result, hence another ensemble member. In this research, the following data-model integration methods have been implemented: standard direct use as model input, meteorological data merging before using as input, EO data infilling based on model results, and data assimilation.

To account for performance differences between ensemble members, overall or dynamic in time (e.g. seasonal or in wetting and drying conditions), weighting methods can be applied (e.g. static weighting, dynamic weighting, model committees). The weighting methods may lead to improved probability distributions of the hydrological simulation outputs. In this research, static and dynamic weighting have been tested.

All the simulation results, using multiple data sources, integration methods, and weighting methods, are wrapped in a hydrological ensemble simulation, which is the final result of the data-model integration framework.

The framework has been tested on a distributed hydrological model of the area of Rijnland in the Netherlands. The hydrological modelling system SIMGRO was used, which is especially suitable for simulating low-land water systems with control structures (weirs, pumping stations), such as the Rijnland water system. Twenty-four ensemble members were constructed with three available land-use maps: LGN6 from the Rijnland Water Board, LANDSAT 5 and GlobCover from satellite products, two soil maps: Dutch database and European database, two observed precipitation data sources: rain gauges and rain radar, and two observed evapotranspiration data sources: weather station and satellite product from Terra/MODIS. The resulting ensemble discharge simulation, the individual members, and the ensemble mean, were compared against the measured discharge from October 2012 to October 2013. The simulations were analysed qualitatively, through hydrographs, and quantitatively using deterministic and probabilistic performance metrics. Deterministic metrics evaluated include Nash-Sutcliffe efficiency (NSE), percent bias (PBIAS), and correlation coefficient. Probabilistic metrics evaluated include Brier score (BS) and relative operating characteristics (ROC).

Visual inspection of the hydrographs showed that the performance of individual members varied in time; hence, no individual member could be identified as uniformly outperforming the others. A sensitivity analysis demonstrated that the parameter uncertainty resulting from different data sources for catchment characteristics, has less effect on the discharge simulation performance than the input uncertainty resulting from different precipitation and evaporation data sources.

Quantitative performance assessment showed that the ensemble mean, with NSE of 0.85 and PBIAS of 13.1%, was more accurate than most individual members in simulating discharge, including the base model (NSE of 0.81 and PBIAS of 22.2%). Although some individual ensemble members performed better for a certain metric over a certain period, overall, using the ensemble mean led to more accurate flow estimates.

Using the complete ensemble as probabilistic simulation (by assigning equal probability to each member) showed good performance for indicating discharge threshold exceedances in Rijnland. High ROC scores (e.g. 0.80 for the 90th percentile discharge threshold) showed that event threshold exceedances can be simulated with high hit rate and low false alarm rate. The ROC results showed an advantage of using the probability distribution of the ensemble simulation, over using individual simulation results. This shows that in spite of the fact that the ensemble size was limited to 24 members, this approach is capable of generating probabilistic simulations of discharge that are more effective in simulating threshold exceedances than deterministic approaches using only a sub-set of the available data sources.

Data assimilation (DA) was applied as data-model integration method for Terra/MODIS actual evapotranspiration. Particle Filter with Residual Resampling DA was used, and this led to improvements in the simulation of discharge from Rijnland over the base model.

Two weighting schemes have been implemented to assess whether the performance of ensemble mean could be further improved. Static weighting based on past performance of individual members, did not lead to an improvement. Dynamic weighting, however, based on time-varying performance, e.g. using previous-day error to give zero weights to the least performing members, did show improvement of ensemble mean with respect to simple averaging.

It is concluded from the Rijnland case study results presented above, that the developed framework for incorporating multiple data sources in hydrological modelling, based on the multi-model ensemble approach, can be applied effectively, improve discharge simulation, and partially account for the parameter and data uncertainty. When combined with the well-known ensemble methods of parameter sampling, including different model types, and forcing with meteorological ensemble forecasts, a next step can be made towards providing reliable hydrological ensemble simulations and predictions to water managers.

Table of Contents

List of figures

List of tables

Chapter 1. Introduction

This chapter introduces the motivation for integrating multiple data sources in hydrological modelling, and explains why ensemble prediction methods were identified as the promising means to achieve this integration. The research objectives are presented, along with the innovation and practical value of this thesis.

1.1 Background

The background of this work is the continuing need for improved hydrological modelling for water management. During the last decades, availability of earth observation (EO) and Numerical Weather Prediction (NWP) data for hydrological modelling and water management has increased significantly, creating a situation that for the same variable, estimates may be available from two or more sources. Precipitation data, for example, can be obtained today from rain gauges, weather radar, satellites, or outputs from numerical weather models. Land-use data can be obtained from land survey, satellite imagery, or a combination of the two.

Each of these data sources provides an estimate of catchment characteristic and related parameters, or of a hydrometeorological variable. Estimates from each data source vary in magnitude or temporal and spatial variability, and have some degree of uncertainty. It is often not possible to convincingly claim that one data source is correct and others are not useful (Beven and Freer 2001; Beven 2006). The use of multiple data sources can capture events that might be overlooked when using only one (Duan et al. 2007). By using several data sources in hydrological modelling, it is expected that the simulation of hydrological variables can be improved (Huffman 1995; Chiang et al. 2007; Yan and Moradkhani 2016).

For a long time, in-situ (ground stations) monitoring has been the main source of hydrological data. The ground stations provide point measurements for various variables, such as water level, precipitation, temperature, wind, etc. Another outcome of in-situ monitoring, is maps of catchment characteristics, such as soil maps, Digital Elevation Models, and land use maps. Such in-situ monitoring is done through field survey and measurements. The ground station monitoring has strengths and limitations. The most notable strength of ground station measurements is the high accuracy, because it is a direct measurement (no interpretation model needed). Nevertheless, in some countries, the quality of the data is sometimes questionable due to the data collection process, involvement of unskilled operators, or due to poor equipment conditions (Michaelides et al. 2009). Most important limitation, however, is that a dense monitoring network is required to get an accurate estimate of the spatial variability, e.g. within a catchment. Purchase costs, man-power required for operation and maintenance, and poor accessibility of remote areas in the catchment, often prohibit installation of a sufficiently dense monitoring network.

Earth observation information for hydrometeorological applications, is continuously getting better, with new satellite and remote sensing technologies, improved resolution and coverage, and advanced algorithms to interpret raw data into catchment characteristics and hydrometeorological variables (Alexandridis et al. 2016). EO data have been used in hydrological studies for a long time. EO is able to capture the details of spatial and temporal processes that may be missed by ground stations (Krajewski et al. 2006). One of the most utilized products is the DEM, for determining watershed characteristics such as catchment delineation and the stream network. Precipitation estimation by radar is also a successful example of the use of remote sensing in hydrology, although it depends on a good algorithmic support to produce reliable data (Berne and Krajewski 2012). Other EO products used in hydrology include land-use, evapotranspiration, soil moisture, surface water extend (including floods), and snow cover maps (see Section 2.4).

Hydrological modelling has been used for many decades to simulate surface- and ground water behaviour, in both quantity and quality. Water managers utilize hydrological models for planning and design, and for predicting water related events, such as floods, droughts. Although it is said that all models are wrong, with a good understanding of hydrological processes in the catchment, a model can be useful to predict what will (forecast) and could happen (scenario) in the catchment. Hydrological modelling has made a vast progress from the rational method in 1850 to physically-based distributed modelling in last decades (Todini 2007; Devia et al. 2015; Salvadore et al. 2015). However, challenges remain, such as the classical problems of defining parameter values, and uncertainty of the model results.

To quantify and take into account uncertainty explicitly in meteorological and hydrological modelling, the field of Ensemble Prediction has been developed (Schaake et al. 2007; Alemu et al. 2010; Strauch et al. 2012). An ensemble prediction represents the uncertainty in the form of multiple estimations. An ensemble can be generated in several ways. A meteorological ensemble prediction system (EPS), for example, re-runs the same meteorological model with every run using slightly different (perturbed) atmospheric initial conditions and parameter values (Buizza et al. 1999; WMO 2012). In a multi-model ensemble, on the other hand, the ensemble is constructed by presenting multiple simulations from different models (Velázquez et al. 2010; Cheng and AghaKouchak 2015). A so-called poor man's ensemble gathers independent model results from several operational centres (Ebert 2001; Cane and Milelli 2010; Perrin et al. 2012).

Despite these advances in hydrological modelling, the multiple data sources available today are not used to their full potential.

1.2 Motivation

So far, the usage of additional data sources in hydrological modelling has mostly been limited to merging two or more data sets for improving one input time series, or to using one additional data source through data assimilation (Loaiza Usuga and Pauwels 2008; Crow et al. 2011; Van Coillie et al. 2011; Liu et al. 2012a; Lievens et al. 2015; Zou et al. 2017; Bai et al. 2018). Using multiple data sources estimating the same catchment characteristics and related model parameters, in combination with multiple data sources for multiple hydrometeorological variables, in one modelling task, is still rare (Xie and Zhang 2010; Yan and Moradkhani 2016; Li et al. 2018). There is no comprehensive framework available to consistently and effectively use multiple data sources in hydrological modelling.

It is hypothesised here that the advances in ensemble prediction methods can help to develop such framework. A multi-model ensemble could be constructed using multiple data sources. A hydrological model can, for example, be build and parameterised with different data sets defining catchment characteristics, such as land-use and soil maps, which would result in different discharge simulations.

It is expected that by smart integration of multiple data sources in multiple model runs, hydrological simulation results can be improved.

1.3 Research objectives

The main objective of this research is to develop and test a data-model integration framework for incorporating multiple data sources in hydrological modelling.

In order to achieve the main objective, several specific objectives are formulated:

1. To develop a methodological framework on the basis of the multi-model ensemble approach, for incorporation of multiple data sources and multiple integration methods, such as data merging and data assimilation, in hydrological modelling
2. To develop and validate a distributed hydrological model for a case study, and to analyse its uncertainty
3. To explore the possibility of using model output to fill-in spatial and temporal gaps in EO data
4. To implement and test an ensemble EO data assimilation scheme
5. To test the data-model integration framework developed, with the case study model and multiple data sources for catchment characteristics and hydrometeorological inputs
6. To analyse the performance of the deterministic simulations resulting from the data-model integration framework (individual ensemble members and ensemble mean)
7. To analyse the performance of the ensemble simulation resulting from the data-model integration framework
8. To test weighting methods to improve the performance of the ensemble simulation

1.4 Innovation and practical value

This research provides a data-model integration framework for the field of hydrological modelling. The innovation lies in the idea to use the ensemble approach to capitalise on all the in-situ and EO data sources available. This allows for more comprehensive data-model integration than has been presented so far.

Next to benefiting research, the framework has a practical value, because hydrological modellers and water managers can use it directly as a guideline to utilise all their available data sources to improve the quality of hydrological model simulations.

The integration framework can be applied in operational decision support systems for planning and design, hydrological forecasting, early warning, and water system control services.

1.5 Terminology

Terminology related to the integration of multiple data sources in hydrological modelling using the ensemble approach, is presented below. Some of the terminology changed over time, depending on the preference of different authors. Hence, it is necessary to have a clear definition for several of the terms that are used in this thesis.

Hydrological model: is a tool to simulate and predict the hydrological variables in a catchment. In this thesis, after the literature review (Chapter 2), with hydrological model we refer to a spatially distributed model, which is used to simulate hydrological processes in the case study of Rijnland (Chapter 4).

Remote sensing: is the acquisition of information about an object or phenomenon without making physical contact with the object and thus in contrast to in-situ observation.

Earth observation: is the gathering of information about planet Earth's physical, chemical and biological systems via remote sensing (RS) technologies supplemented by earth surveying techniques.

Natural catchment: is a catchment where in general the water is able to flow naturally with little to no effect of human influence.

Controlled water system: is a catchment or water system in which the water flow is regulated by human-made structures. Hence, the hydrological variables and/or system states are strongly influenced by control structures, resulting in a less natural system. This relates to the case study of Rijnland (Chapter 4), which is a land-reclamation area controlled by an irrigation and drainage system.

Multi-model ensemble: is an ensemble formed by several models that simulate the same output variable. In this thesis, the ensemble is generated by using multiple data sources for catchment characteristics and hydrometeorological variables to parameterise and drive the hydrological model. Although in this study the same modelling software is used, and model topology is constant, the different data sources are affecting model parameter values and forcing, hence the simulation results come from different model instantiations.

Validation: is testing the model results against observed data for a period other than the calibration period. The aim is to test the robustness of the model and its ability to mimic the hydrological response of the catchment.

Mutual validation: is a two-way validation, i.e. the correctness of both modelled and observed data is doubted.

1.6 Thesis outline

After the introduction presented above, the thesis outline is as follows.

Chapter 2 presents the literature review of work done around the integration of different data sources in hydrological modelling. The review begins with available data sources in hydrometeorology and their usage. Then, the role of hydrological modelling and uncertainty is analysed, followed by exploring data-model integration research. Ensemble prediction methods, including the multi-model ensemble, are discussed in the last section.

Chapter 3 presents the methodology of the research, i.e. development of the data-model integration framework based on the multi-model ensemble approach, applying multiple data-model integration methods, and performance evaluation.

Chapter 4 presents and analyses the catchment characteristics and hydrometeorological data sources available for the case study. The data sources consist of in-situ measurements and earth observation estimates, and represent land-use, soil type, precipitation, evapotranspiration, soil moisture, water level, and discharge.

Chapter 5 presents and discusses the model development for the case study. The model was validated against observations of the result variable, which is discharge through the main outlets. However, there are several other observational time series available in the case study area, hence the model results were further validated against these, i.e. local drainage discharge, local surface water level, ground water level, evapotranspiration, and soil moisture.

Chapter 6 presents tests of data-model integration methods and analyses their effect. First method tested is simple direct use of every data source to parameterise and drive a hydrological model. Secondly, combining data sources into a merged time series and feeding it to the hydrological model is presented. The third data-model integration method presented is to use the model's output to improve an input data source and feed it back to the hydrological model. The last method presented is to use an additional data set to update the model during simulation: data assimilation. In this research, an ensemble particle filter method is implemented.

Chapter 7 presents the integration of multiple data sources into a multi-model ensemble. In this chapter, each of the available data sources, in different combinations, is used to parameterise and as input to a hydrological model to form an ensemble of discharge simulations that represents the combined strengths and weaknesses of the input data. Discharge simulations from the individual ensemble members, ensemble-mean, and weighted ensemble are compared. Apart from the performance criteria for a deterministic, single, simulation, ensemble simulation performance is analysed with the Brier score and relative operating characteristics.

Chapter 8 demonstrates the data-model integration framework in an operational system.

Chapter 9 summarizes the conclusions and recommendations on integrating multiple data sources in hydrological modelling as presented in this thesis.

Chapter 2. Literature review

This chapter reviews data sources and their integration in hydrological models. The chapter begins with available data sources in hydrometeorology and their usage. Then, hydrological models and uncertainty are reviewed with a focus on the case study model, followed by the data-model integration research. The main idea in this research for developing a new framework for using multiple data sources and data-model integration methods in hydrological modelling is to build on the ensemble prediction approach, which is discussed in the last section of this chapter.

2.1 Sources of hydrometeorological data

2.1.1Ground station data

In-situ measurement data have been used as a main data source in hydrology for a long time. It serves high reliability and high accuracy. Data gathering techniques also have been improved greatly with the use of automated stations, wireless communication, and centralized computer-based data processing.

However, the in-situ measurement lacks spatial properties due to the fact that it measures only one point in space. For some hydrological variables spatial properties are not important, e.g. if one is interested only in the discharge at a catchment outlet. On the other hand, variables such as rainfall and vegetation cover need detailed spatial representation.

There are several possibilities to increase spatial representation of point measurement data. One way is to increase the point measurement density in the area, but this leads to high costs in building and operating measurement stations, which is a big problem in developing countries. Another way to obtain spatial properties is to interpolate between points. Various spatial interpolation methods have been developed, such as traditional deterministic distance-based methods (e.g. Inverse Distance Weighting (IDW)), stochastic variance interpolation methods (e.g. Kriging), and data-driven interpolation methods, e.g. Artificial Neural Networks (Teegavarapu and Chandramouli 2005; Seo et al. 2015; Kumari et al. 2017).

Even though in-situ measurement values are generally reliable, they are still subject to various errors. Rain gauges, for example, suffer from environmental influences, such as wind, evaporation, and raindrop size temporal and spatial variation (Michaelides et al. 2009). There are also structural errors in rain gauges, e.g. water loss during measurement, raindrop splash out, and adhesion losses.

Another problem common to in-situ measurements, is missing data, e.g. due to problems in measurement tools, blockage by external influences (e.g. tree growth), and human errors in manual readings. The filling of missing data is an important step in hydrological modelling. The filling of missing data by using surrounding stations can be achieved by the same methods as referred to in the previous paragraph for spatial interpolation.

2.1.2 Earth observation data

Earth observation (EO) data are acquired from remote sensing equipment, normally from satellite and airborne vehicles. EO data are obtained through measurements of the electromagnetic spectrum to obtain properties and characteristics of an area (Schultz and Engman 2000). The electromagnetic spectrum can be a reflection from a source or emitted directly from the earth surface.

Representation of spatial variability is the biggest advantage of EO data over ground-based point measurements. The data also can have a good temporal coverage if the satellite re-visit time is short, which is common in today's satellites. The spatial variability is important for spatially distributed models, and with denser temporal availability, the EO data is able to drive real-time simulation (Stisen et al. 2008). Berne and Krajewski (2012) also stated that ground based measurements may not provide reliable data during extreme events, such as flash floods, while remote sensing data, e.g. from radar, are well suited for modelling such events. Furthermore, Krajewski et al. (2006) reiterate about the ability of EO to capture the details of spatial and temporal processes, which are the main causes of complexity and heterogeneity in hydrologic interactions.

However, there are some disadvantages of using EO data. For example, EO never measures object's properties directly, instead it interprets the electromagnetic spectrum into required properties. Furthermore, a dense cloud cover can reduce the usability of satellite images, especially those in the parts of the electromagnetic spectrum that are unable to penetrate clouds, e.g. the visible spectrum and infrared spectrum. The microwave spectrum, however, is able to provide better cloud penetration (Blyth 1993). The indirect measurements of remote sensing lead to another disadvantage: the need of good interpretation methods and algorithms in order to produce reliable data (Schmugge et al. 2002; Cherif et al. 2015). Furthermore, the best methodology of interpreting and calibrating raw satellite data could be different for each geographic area and depends on its characteristics. The indirect measurements also lead to higher uncertainty as stated by Berne and Krajewski (2012). Despite advances in spatial and temporal resolution, the resolution may be insufficient for a particular hydrological modelling application, in which case downscaling is required (Lanza et al. 1997; Atkinson 2012).

For hydrological models one of the main EO products used is the precipitation map, usually provided by radar or satellite, with integration (e.g. bias correction) of ground station and meteorological model data (Michaelides et al. 2009). Other hydrometeorological fluxes that can be obtained from satellite imagery include evapotranspiration and snowmelt. Hydrometeorological states include land-surface temperature, near-surface soil moisture, vegetation cover, snow cover, snow-water equivalent, water quality, and landscape roughness (Schmugge et al. 2002). Moreover, EO also can provide land-use and catchment characteristics, surface water states, soil erosion monitoring and ground water recharge (Schultz and Engman 2000; Hartanto et al. 2015).

Several researches have been conducted to improve the estimates from EO observations. Artificial Neural Networks can be used to improve the estimates of precipitation from satellite imagery (Evora and Coulibaly 2009). Soil moisture from EO can be used to improve the accuracy of satellite-based rainfall products (Crow and Bolten 2007; Crow et al. 2011). Li and Shao (2010) applied statistical methods in merging satellite derived precipitation data with ground based rain gauge data, and achieved better rainfall estimations.

Out of the many available EO hydrometeorological data products, five products with high potential benefit for hydrological modelling have been used in this research: land-use/land cover, leaf area index, actual evaporation, soil moisture, and precipitation. These products are described in detail in Chapter 4.

2.1.3 Numerical Weather Prediction

The term "Numerical Weather Prediction" refers to application of computer models of atmospheric processes and ocean dynamics to predictions of weather conditions. Global and regional models simulate weather in many regions in the world. There is distinction between short-term, mid-term, and long-term weather forecasts; short-term concerns weather predictions for the coming hours or days, while the long-term is used, for example, for climate change analysis.

Weather forecasting has moved from deterministic forecasts to probabilistic forecasts. Deterministic forecast uses a numerical model to represent dynamic process in the atmosphere using physical laws and to make one prediction of future weather conditions. In probabilistic weather forecasting, based on uncertainty information, the probability distribution of upcoming events is presented, and one of the examples is ensemble prediction system (EPS).

In ensemble prediction systems, several initial conditions of the atmosphere are fed into a numerical model, such that a number of predicted events can be simulated. The atmospheric initial conditions for the first run are taken from the deterministic forecast. The initial conditions are then perturbed with the aim to have the resulting forecasts equally likely to happen (van Andel et al. 2014). The ensemble methods that introduce small differences in initial conditions are usually called the Perturbed Initial Conditions (PIC) ensemble. Another ensemble-based approach is to run different models, each with different key parameters; this strategy is called Perturbed Physics (PP) ensemble (Tapiador et al. 2012). Ensemble meteorological forecasts take into account the chaotic behaviour of the atmosphere. Each model run is called an ensemble member, and the resulting time series, e.g. for temperature and precipitation, can be extracted for each member and for each grid cell, and used as input to force hydrological models. Both in meteorological and hydrological ensemble forecasting, in most cases, biases in the forecast probability distributions local to the particular catchment do exist. These biases, e.g. in ensemble mean and spread, can be reduced by post-processing methods (Verkade et al. 2013 ; Zalachori et al. 2012).

2.2 Hydrological models

Hydrological models have long been used to help water managers to support their decision-making, e.g. on decisions related to flood forecasting. By trying to represent hydrological processes of a water system in mathematical equations, hydrological models help water managers to understand the system and to predict what will happen next.

Mathematical hydrological modelling begins back in 1850 when the Rational Method was introduced by Mulvany, with the use of the relationship between time of concentration and peak flow (Todini 2007). Later on, more physically meaningful models emerged, trying to represent real-world processes with complex mathematical equations. However, due to the limitation of resources and data, in the 1960s a simple lumped model with interconnected conceptual elements was considered as the best representation that could be achieved. At the end of the 1970s, a new type of lumped physically based hydrological model was developed, based on the assumption that the hydrological processes are mainly determined by dynamic processes of saturated areas. The models assumed that all precipitation goes into the soil and after saturation of upper soil layer surface runoff develops (Todini 2007). Physically based spatially distributed models also began to develop, based on full dynamic equations, with complex calculations in each grid cell, trying to represent the real world as close as possible. This concept was applied, for example, in the SHE model and evolved into a robust physically-based spatially distributed hydrological model further developed by DHI and known as MIKE-SHE (Abbott et al. 1986). However, spatially distributed modelling requires a lot of data and high computational time, so simplified physically-based spatially distributed hydrological models were introduced later, such as LISFLOOD and WATFLOOD; they use simplified equations and have lighter computational load (Todini 2007). Although spatially distributed models seem to be very close to the real world representation, they still suffer from a number of issues: nonlinearity, scale, equifinality, uniqueness and uncertainty (Beven 2001). Furthermore, Beven (1989) stated that spatially distributed models could still have the same disadvantages as lumped models, such as error in estimation of parameters and variables. More recently, indeed it was indicated that informed estimation of hydrological model parameter values, e.g. on the basis of observations and process understanding, is still a key challenge (Clark et al. 2017).

Data driven (statistical) models characterize the connection between input and output, without explicit formulation and calculation of the underlying physical processes. Using computational intelligence and machine learning techniques, data driven models can be seen as important class of models complementing the traditional process- (physically-) based models (Solomatine 2006). Many forms of data driven models have been utilized in hydrological modelling, which can be divided into two approaches, linear and non-linear. Linear models such as Auto Regressive (AR), Auto Regressive Moving Average (ARMA), and Auto Regressive Integrated Moving Average (ARIMA) models, are the simplest ones, and are based on assumptions of stochasticity of a single-variate time series, and have been used e.g. in river discharge prediction. Multi-linear regression models link several variables in a linear equation. Non-linear regression techniques are also well-developed, and include, for example, K-Nearest-Neighbours (KNN) algorithm, Artificial Neural Network (ANN) and Support Vector Machine (SVM) (Wu and Chau 2010). Wu and Chau (2010) pointed out that determining the best data driven model for a catchment is difficult, because it is highly dependent on catchment characteristics, prediction length, and whether local or global approximation technique can be used.

Clark et al. (2017) call for an effective use of the different hydrological model types available as described above, to increase understanding of hydrological processes and their representation.

2.2.1 Models of controlled water systems

The case study of this thesis is a low-lying catchment in the Netherlands with a controlled water system (Chapter 4). This sub-section describes the particular challenges when modelling such system, and the following sub-section describes the modelling system selected for the case study, SIMGRO.

A controlled water system is a system where variables and/or system states are set by control structures in addition to natural processes (van Andel et al. 2010). A typical example of a controlled water system is an irrigation system, with weirs, channels and gates to regulate water flow.

Inside the controlled water system, human influences highly affect the hydrological processes. For example, during a heavy rain, the water is pumped out from the system to avoid a flood, which would occur in a natural system. Sometimes the human influences are going beyond that, when pre-pumping, before the event, determined by forecasted heavy rainfall. Another example is controlled flushing of a system to maintain good water quality.

Modelling a controlled water system is different from modelling a natural water system (van Andel et al. 2010). Modelling a controlled water system is often challenged with a high degree of freedom, e.g. modelling too much reservoir release can compensate for simulating too much inflow resulting in the 'correct' reservoir level, and with unpredictable records in the observed data caused by unknown events that altered the control action. A pumping station, for example, could have the capacity reduced, or a pre-pumping decision could be taken by the operator. However, for a controlled water system there is usually more data available than for a natural system. The control structures often record discharge and water levels up- and downstream, and there usually is information on canal and weir dimensions.

SIMGRO (van Walsum and Veldhuizen 2011) is a distributed hydrological modelling system that can incorporate the vast availability of water system information, and is especially suitable for low-lying irrigation and drainage systems (land reclamation areas: polders) in the Netherlands, such as the case study of this research: Rijnland.

2.2.2SIMGRO modelling system

SIMGRO is a modelling hub of three different models, a soil-water-atmosphere transfer model, a surface water model, and a ground water model. The integration of these three models is done through the exchange of shared states. The ground water level and recharge, for example, are shared between the ground water model and unsaturated zone in soil-water-atmosphere transfer model (Van Walsum et al. 2011).

The soil-water-atmosphere model utilized in SIMGRO is MetaSWAP, which handles the processes in the unsaturated zone and water transfer from/to the atmosphere. MODFLOW model is responsible for calculating ground water processes in three dimensions. The surface water flow is calculated by SurfW model, a simple storage basin model. Although the MetaSWAP model only calculates vertical flow, with the connections to the 1D surface water, and the 3D ground water model, a spatially distributed hydrological model calculation can be achieved.

SIMGRO is strong in modelling shallow water level and water systems with pumping stations. The software is able to use target water levels as an input, which is common in polder water systems in the Netherlands.

Figure 2-1 presents the hydrological processes that are simulated in the SIMGRO model. In addition to natural hydrological processes, the human interference e.g. sprinkling, ground water extractions, and sub-surface irrigation can be modelled. Furthermore, the pumps and weir operations are modelled in the surface water model. The water input to the model may come from precipitation, ground water and surface water, while the outflow may come from interception, evapotranspiration, ground water outflow and surface water outflow.

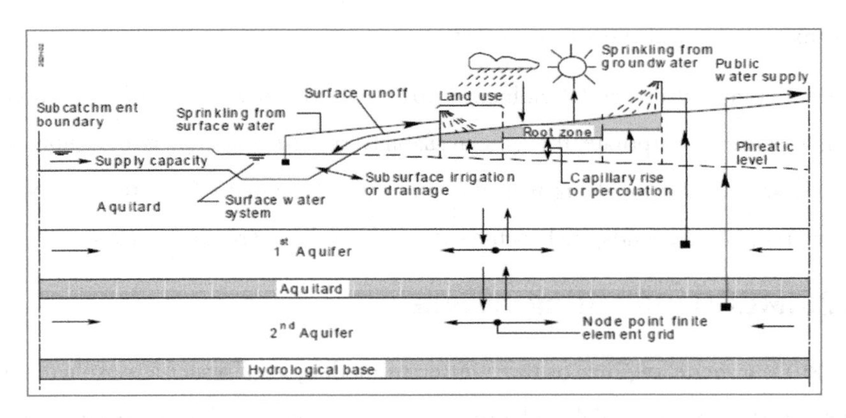

Figure 2-1. Hydrological processes that are modelled in SIMGRO (Van Walsum 2011)

ET_a calculation in the hydrological model is one of the important modelling processes for this thesis. ET_a calculation in SIMGRO uses the Makkink method (de Bruin and Lablans 1998), which is a commonly adopted method in modelling hydrological processes in the Netherlands. However, De Bruin and Lablans also give a remark that the method might incorrectly estimate ET_a in a winter period when the radiation is not the main driving force. In order to calculate the potential evaporation (ET_p), the Makkink reference evapotranspiration (ETr_{mak}) is multiplied by an evaporation factor (f). The evaporation factor is correlated to vegetation and/or land-use types.

$$ETp = ETr_{mak} \times f \tag{2.1}$$

The evaporation factor (f) consist of f_t for transpiration, f_{Eic} for interception, f_{Ebs} for bare soil, and f_{Epd} for ponding water evaporation. Furthermore, there are several calculations following the ET_p calculations e.g. the temporal fraction, LAI, and extinction coefficient of solar radiation. SIMGRO model needs input that contains the information of the crop growth to better simulate the evapotranspiration. In addition, a coupling to WOFOS (Diepen et al. 1989) crop growth model is available.

The limitation of vegetation root uptake is determined by the water pressure in the root zone, using the ET_a limitation method by Feddes (1978). The limitation is represented by a soil moisture reduction factor a_E which is a reduction factor to potential evaporation, as expressed in equation 2.2.

$$ETa = ETp \times \alpha_E \tag{2.2}$$

The full model descriptions and theory of evapotranspiration in SIMGRO can be seen in the SIMGRO theory and model implementation (Van Walsum et al. 2011).

There are two kinds of processes in SIMGRO, the fast processes of surface water and unsaturated zone, and the slow processes of ground water flow. Unsaturated zone in SIMGRO is simulated as parallel vertical columns, each connected to a cell in the MODFLOW simulation. The phreatic surface acts as a moving boundary between unsaturated zones modelled by MetaSWAP and saturated zone by MODFLOW model. As a result, all ground water lateral flow is simulated by the MODFLOW model where three-dimensional flow occurs.

The relationship between a MetaSWAP column and the surface water, occurs via two paths. One path is over the soil surface, i.e. run-off and run-on, and the other is through subsoil i.e. drainage and infiltration. The drainage from the soil to the surface water is a gravity flow involving head differences. The same process applies to the recharge from unsaturated zone to the ground water with a diffuse process. The flow is reduced by several resistance coefficients depending on the soil type and canal shape. The flow interaction of ground water with surface water can be simulated in two ways, by using the standard MODFLOW drainage and river module, or by using the SIMGRO module.

The SurfW model can handle weirs and pump operations. The weir simulation is based on the target water level in the surface storage. When the level is reached, water will flow to another SurfW unit. The flow intensity is determined by a Q-h relationship. The handling of pumps is similar to the weir, with no backflow. The pumps will operate when the target water level is reached, at a rate according to their pumping capacity. Different target water levels for winter and summer periods can be defined in SurfW model, which is relevant for the Rijnland case study (Chapter 4), where seasonal target levels are applied. The SIMGRO model development for the Rijnland case study is presented in Chapter 5.

2.3 Uncertainty in hydrological modelling

Identifying and quantifying sources of uncertainty is a major goal in environmental modelling, including hydrological modelling. The identification covers the following areas: model inputs, boundary and initial conditions, model parameters, model representations of physical processes, model numerical formulation, and observations of the system behaviour (Gourley and Vieux 2006). Failure in identifying sources of uncertainty, will lead to difficulties in model calibration and parameters estimation, which consequently gives uncertain output, especially in forecast output (Vrugt et al. 2005).

Quantifying uncertainty is the next important step in uncertainty assessment, and considering each source separately will determine which sources have significant impact on uncertainty of the final output.

There are several methods available to assess uncertainty in hydrological modelling. Uncertainty assessment methods that focus on parameter uncertainties including classical Bayesian techniques, the pseudo-Bayesian, set theory, and recursive model and parameter identification techniques (Mantovan and Todini 2006). Generalized Likelihood Uncertainty Estimation (GLUE) (Beven and Freer 2001; Beven and Binley 2014) is a pseudo-Bayesian method widely used. Moradkhani et al. (2005) proposed particle filtering as a sequential Bayesian filtering to assess uncertainty in model states and parameter values.

However, those methods typically account for parameter uncertainty (Mantovan and Todini 2006), putting aside uncertainties from input data and model structure. Analysing uncertainty in model structure can improve overall performance of hydrological simulations (Butts et al. 2004). Input and model structure uncertainty can be merged with parametric uncertainty to assess the final uncertainty. A method proposed by Vrugt et al. (2005) called Simultaneous Optimization and Data Assimilation (SODA) addresses input, output, model structure and parameters uncertainty. Additionally, Shrestha and Solomatine (2008) presented an approach termed UNEEC, where data driven techniques can predict residual uncertainty in calibrated models (see also Dogulu et al. 2015). Shrestha et al. (2014) developed MLUE technique in which the results of a GLUE analysis are used to train a machine learning model able to predict model output uncertainty in real-time.

2.4 Integration of data and models

In this research, data-model integration is understood as any combination of a data source or multiple data sources with a numerical model. For example, merging of in-situ rainfall measurements with radar rainfall estimates to come to a more reliable rainfall estimate to be used as input in a hydrological model, is considered a data-model integration method. This section explores earlier research on integration of information sources and hydrological models and the data-model integration methods used therein.

2.4.1 Integration of sources of information in hydrological models

Figure 2-2 shows for each source of information (in-situ measurements, EO, numerical models) the variables concerned, and the relationship with other information sources and the hydrological model. In the following paragraphs examples of research exploring these relationships are given.

The most obvious use of in-situ meteorological data (top-right corner of Figure 2-2), in-situ water level time series (top-left corner of Figure 2-2) and water management data (bottom right corner of Figure 2-2) with the hydrological model, is through calibration and validation. Calibration of hydrological models is based on comparing model result with in-situ data at catchment outlet such as water level and discharge (Andersen et al. 2001; Sahoo et al. 2006; Jian et al. 2017) or combining it with other in-situ measurements like soil moisture and infiltration (Loaiza Usuga and Pauwels 2008). In-situ measurements of output variables, e.g. streamflow, have also been integrated through data assimilation (Aubert et al. 2003; Liu et al. 2012a; Mazzoleni et al. 2015). The catchment characteristics and water system data, such as soil maps, land-use maps, channel dimensions, and slopes, are mostly used to build and parameterise the hydrological model.

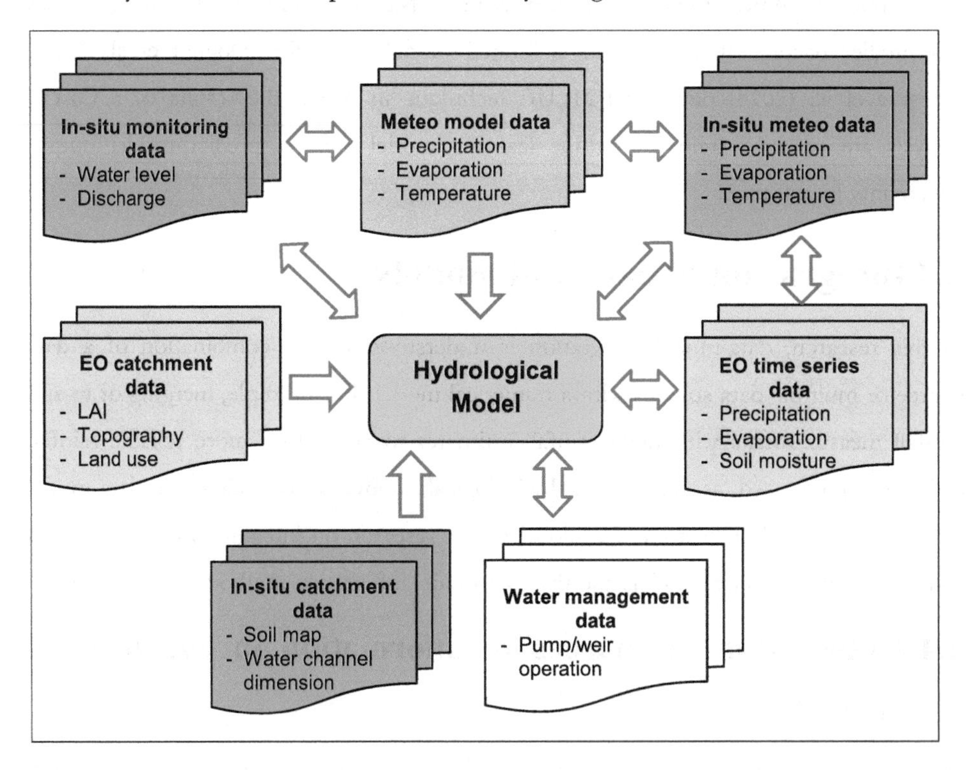

Figure 2-2. Sources of information available for integration with a hydrological model

Integration of in-situ monitoring data with EO data also has been studied by many researchers (Sun et al. 2000; Sokol 2003; Haberlandt 2007; Cole and Moore 2008; Crochet 2009; Li and Shao 2010; Jiang et al. 2012). The studies are mainly revolving around precipitation data correction between radar data and ground station data. However, with the advance of remote sensing technology and hydrological interpretation algorithms, more satellite-based data sources and in-situ data can be integrated.

Numerical Weather Prediction models integrate in-situ and EO data mainly through data assimilation (Sokol 2011; Jiang et al. 2012). The integration is meant to correct the model states operationally (in real-time), hence a better output can be obtained right after new in-situ monitoring data is available (Einfalt et al. 2004). Feeding the latest measured data into the model is important in meteorological forecast models because of the strong influence of initial conditions on forecast results.

Crow and Bolten (2007) showed that satellite-based soil moisture data, can be used to estimate and correct rainfall data products. The Kalman Filter method that they use is able to produce results comparable with ground station data in heavily gauged areas in the United States. In further research, they improve the algorithm and implement it in less gauged areas in Africa (Crow et al. 2011).

Integration of EO data with hydrological models also has been done by many researchers. There are various possible variables to be incorporated, such as land surface temperature, near-surface soil moisture, vegetation cover, snow cover, snow-water equivalent, and water quality (Schmugge et al. 2002). In addition, EO also can provide more general data such as land-use and soil erosion monitoring (Schultz and Engman 2000; Hiep et al. 2018). Generally, EO information is used in semi-spatially distributed and fully spatially distributed models. However, Pauwels et al. (2001) pointed out that it can also be beneficial to assimilate EO data in lumped hydrological models, e.g. for soil moisture.

Many researchers are working on assimilation of EO data with land surface models to improve accuracy of hydrological model simulations. One example is using remote sensed evapotranspiration estimates to calibrate and validate the evapotranspiration computed by hydrological models (Schuurmans et al. 2003; Immerzeel and Droogers 2008; Rientjes et al. 2013; Zou et al. 2017). Another option is to compare EO-based leaf area index (LAI) with hydrological model results (Boegh et al. 2004; Chen et al. 2005; Stisen et al. 2008). Andersen (2002) tried to combine LAI map and precipitation data from satellites into hydrological modelling. Although the LAI data improved the result, combination with precipitation data gave no further improvement compared to ground based rainfall data. LAI data also can be integrated into conceptual rainfall-runoff models as pointed out by Li et al. (2009). Integration of vegetation growth model with LAI maps is done by Curnel (2011) and Pauwels et al. (2007) through data assimilation.

Some researchers attempted to validate soil moisture from hydrological model with satellite-based estimates (Jackson et al. 1996; Reichle et al. 2002). Reichle et al. (2002) applied data assimilation of soil moisture data from EO into a land surface model. Data assimilation of soil moisture has been applied by many researchers, e.g. Margulis et al. (2002), Lee et al. (2011), Han et al. (2012), Lievens et al. (2015). The combination of soil moisture with stream flow DA has successfully been carried out by Wanders et al. (2014), Yan and Moradkhani (2016) and Li et al. (2018).

Clark (2006), Nagler et al. (2008), and De Lannoy et al. (2012) tried to integrate snow cover area into a hydrological model by using data assimilation. Snow cover data assimilation from EO into a macro scale hydrological model also gave a good result as pointed out by Andreadis and Lettenmaier (2006). While Moradkhani (2008) indicated that adding soil moisture data assimilation together with snow cover area is possible and could give a better result. Data assimilation of EO ET_a, although less extensive as other EO data, has been researched by, for example, Schuurmans et al. (2003), Olioso et al. (2005), Qin et al. (2008), Vazifedoust et al. (2009), Irmak and Kamble (2009), and Zou et al. (2017).

Most of the integration studies that have been done, show improvement over the single data set or of the hydrological model output.

2.4.2 Integrating multiple sources of information

The previous section shows that a lot of research has been done on integration of hydrometeorological information from different sources, however, most of these researches focus on integration of only two information sources (Andersen et al. 2001; Aubert et al. 2003; Sahoo et al. 2006; Liu and Gupta 2007; Loaiza Usuga and Pauwels 2008; Lee et al. 2011; Liu et al. 2012a; Bai et al. 2018). Some DA studies assimilated two data sources in a hydrological model (Wanders et al. 2014; Yan and Moradkhani 2016; Li et al. 2018), but integration of more than two information sources for multiple variables is, to our knowledge, still rare. Reichle (2008) already stated that there is a need to develop multi-variate approaches to merge and assimilate several data types as part of the same system. Xie and Zhang (2010) have tried to add multiple data types into hydrological modelling through data assimilation, although they worked partly with synthetic data (generated by SWAT model). The researchers added (real) soil moisture and evapotranspiration data from remote sensing into their base-line hydrological model, which already had runoff (synthetic) data assimilation. The research concluded that by adding several data types in the assimilation process, the model results could be augmented.

2.5 Ensemble Prediction

The uncertainty in hydrometeorological forecasting is often taken into account by using probabilistic prediction methods. The probabilistic prediction gives several prediction traces or a probability distribution, instead of one deterministic prediction. The many predictions or probability distribution come from the uncertainty in parameter values, inputs, and model structure. One of the probabilistic forecasting methods is Ensemble Prediction, where a set of predicted events is presented: the ensemble. Ensemble prediction has been utilized in meteorological forecasting for decades and is accepted as a good method to simulate atmospheric variables (Cheng and AghaKouchak 2015; Pappenberger et al. 2015). For example, in the case of rainfall prediction, the ensemble can be generated by perturbation of the atmospheric initial conditions of the model. As a consequence, each model run results in a different rainfall prediction (van Andel et al. 2014; Skinner et al. 2015). This type of ensemble is called ensemble prediction system. In another ensemble simulation method, called poor man's ensemble, the ensemble is generated by selecting several available predictions from different providers. This method can sample uncertainties from different sources and is less prone to systematic errors and biases (Ebert 2001). The meteorological ensemble products have been used as input in hydrological modelling, especially the ensemble of precipitation data (Chiang et al. 2007; Strauch et al. 2012; van Andel et al. 2014; Newman et al. 2015; Skinner et al. 2015; Thiboult et al. 2016).

Another approach is to run different models to create the members, that is to build a multi-model ensemble. With several models simulating the same hydrological variable, the uncertainty in parameters and model structures can be accounted for (Georgakakos et al. 2004). The multi-model simulation has been applied and evaluated in hydrology (Ajami et al. 2007; Duan et al. 2007; Velázquez et al. 2010; Perrin et al. 2012). The multi-model ensemble has shown improved performance over deterministic simulation.

The multi-model ensemble outputs can be of course combined into a single output. The simplest way is to apply an equal weight to each of the members and produce one estimation of the model result variable. More complex weighting methods are also available, such as artificial neural network (ANN), Bayesian model averaging (BMA) or linear regression (Shamseldin et al. 1997; Ajami et al. 2006).

Such weighting also can be applied dynamically. One of the examples is to build different models that can be calibrated to different parts of the hydrograph (e.g. high flow and low flow), i.e. to form a model committee. This provides a weight that change throughout time (Oudin et al. 2006). Fenicia et al. (2007) used a so-called fuzzy committee approach to also combine two specialized models calibrated differently for low and high flows. The further extension of this dynamic weighting method for model committees, which applies switching between model results based on the upcoming input variable or climate conditions, was presented by Kayastha et al. (2013).

With the availability of data from different sources for the same catchment characteristics and hydrometeorological variables, another type of multi-model ensemble could be created, using different combinations of data sources to construct, parameterise and drive the hydrological model. By using different sources of data, each of the resulting simulations can be treated as originating from a different model. The different data sources might give different parameter values or structure to a model. Different soil types have different soil parameters, which lead to a different hydrological model simulation. Based on this understanding, the ensemble that is generated from different data sources can be considered a multi-model ensemble simulation. In hydrological ensemble modelling studies, the ensemble is mostly created by using ensemble meteorological input and by random sampling from the hydrological model parameter space (van Andel et al. 2008; Golembesky et al. 2009; Alemu et al. 2010; Jaun et al. 2011; Regonda et al. 2011;

Liao et al. 2014). Creating the ensemble simulation from multiple data sources for the same catchment characteristics in conjunction with multiple data sources for the same observed hydrometeorological variables has not been researched much.

2.6 Summary and gaps in research

Increasingly, alternative information sources describing the same components of the water system become available for hydrological modelling. A methodology is needed to utilise this abundance of data as effective as possible. From the literature review above, it can be seen that, while extensive research has been carried out on integrating a certain new data source available, or on improving a particular data-model integration method, integration of multiple data sources is still limited. A comprehensive approach for using multiple data sources and multiple data-model integration methods is not available.

One method that we found worth exploring, and which has the potential to integrate multiple available data sources, is the multi-model ensemble approach. It is this approach that is chosen to be the central point to develop the methodological framework of this research.

Chapter 3. Methodological framework

This chapter presents the data-model integration framework. The multi-model ensemble approach was selected as the basis for integration, where each of the members is built from using different data sources. Data-model integration methods, e.g. data assimilation, can be selected and applied to construct additional members. All the simulation results, using multiple data sources and integration methods, are wrapped in a multi-model ensemble. Weighting schemes can be applied to derive the probability distribution and ensemble mean.

3.1 Introduction

In the data-model integration framework outlined below, each data source is used to derive catchment parameter values or hydrometeorological input time series for a hydrological model. Each unique combination of catchment characteristics and input data sources thus leads to a different hydrological simulation result - a new ensemble member. Together, the members form an ensemble of hydrological simulations. By following this approach, all available data sources are used effectively, and their information is preserved, resulting in a hydrological ensemble simulation that quantifies part of the data and parameter uncertainty.

3.2 Multi-model ensemble approach

The framework consists of three main stages (Figure 3-1), concerning Data, Modelling, and Ensemble simulation analyses respectively.

In the first stage, the available data need to be quality-checked, and corrected or rejected if necessary (Data sources selection box in Figure 3-1). The quality assessment takes into account such factors as data gaps, physically impossible values, constant values, values above thresholds, and consistency with other observed time series.

In the Modelling stage (centre of Figure 3-1) model building and parameterisation takes place using the multiple available data sources for the same catchment characteristics, such as for land use and soil type. The instantiated models are forced with the multiple available hydrometeorological input data sources, such as for precipitation and evaporation.

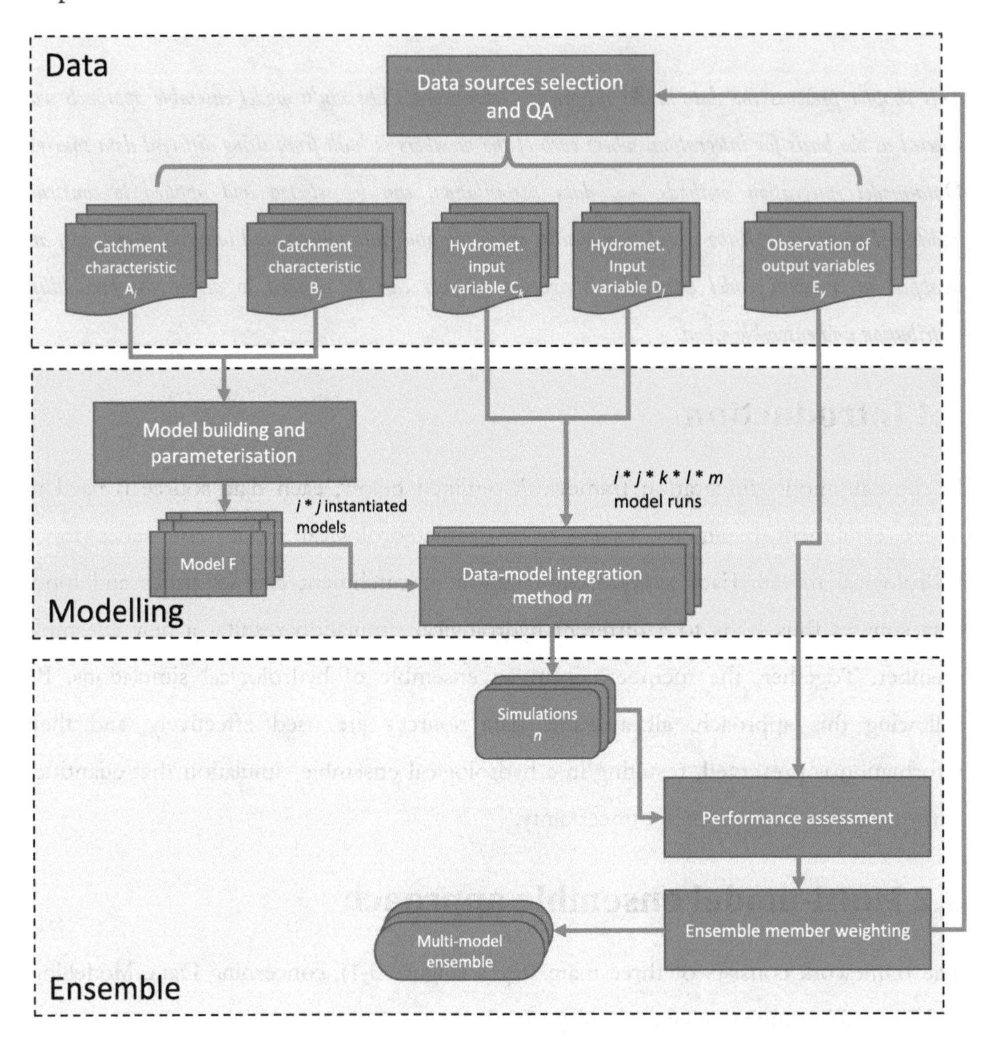

Figure 3-1. Framework for integration of multiple data sources in hydrological modelling based on the multi-model ensemble approach

Not all EO and NWP products can be used directly to feed the input into a hydrological model. A model may, for example, require reference evapotranspiration as input, while the satellite-based product provides actual evapotranspiration. Methods for integrating the data sources with the hydrological model, such as Data Assimilation, have to be applied (Data-model integration box in Figure 3-1). These methods also include techniques to improve the input data, e.g. through merging and through data infilling or correction using model outputs. The data-model integration methods applied in this research are listed in Section 3.3. Each alternative method of data-model integration leads to yet another unique hydrological simulation result, hence another ensemble member. In this way the framework provides a consistent way of incorporating several available integration methods for one hydrological modelling task.

In the Ensemble stage, all the simulation results, using multiple data sources and data-model integration methods, are wrapped in a multi-model ensemble simulation and the performance is assessed (lower-right box in Figure 3-1). Performance is assessed by comparing simulation results with observations of the output variables (upper-right box in Figure 3-1). To account for performance differences between ensemble members, weighting methods can be applied (e.g. static weighting, dynamic weighting, or model committees accounting for a spectrum of hydrometeorological conditions). The weighting methods applied in this thesis are described in Section 3.4. The weights of the ensemble members determine the probability distribution of the resulting multi-model ensemble simulation, and derivatives such as the ensemble mean.

The performance assessment and weighting results may lead to changes in the selection of data, additional input data correction, or other updates in the earlier steps of the framework, allowing for an iterative process to improve the ensemble hydrological modelling results.

3.3 Data-model integration methods

Inside the Modelling stage, any data-model integration method can be applied (centre box in Figure 3-1). In this research, four data-model integration methods are applied.

Data-model integration method 1 concerns the direct use of forcing data as input to run the model.

Data-model integration method 2 concerns the merging of input data sources, i.e. merging of precipitation data from ground station monitoring and from weather radar.

Data-model integration method 3 concerns going back to input remote sensing data interpretation with the aim to improve EO data if a systematic discrepancy between model output and observations is identified. Next to improving final hydrological simulations, this feedback loop provides the possibility of critical assessment and improvement of the remote sensing data product by the EO data provider.

Data-model integration method 4 concerns data assimilation. In this thesis input-updating is applied to match modelled ETa and EO ETa, using the particle filter method.

The methods and the way they are applied are described in detail in Chapter 6 on Data-model integration.

3.4 Weighting methods

The third and last stage of the developed data-model integration framework is to analyse the set of hydrological simulations as a multi-model ensemble. Performance assessment by using available observations of resulting time series needs to be done (right box in the Ensemble stage of Figure 3.1). The performance assessment can be used to identify possibilities for improvement in the earlier steps of the modelling process, and it can be used to set the weights of the members, such that the probability distributions of the ensemble can be derived and optimised (lower right box in the Ensemble stage of Figure 3.1).

In this work we are considering linear combination of outputs (weighting), which however may be dynamic, i.e. weights may change in time, e.g. they may depend on model members performance, or on the hydrometeorological conditions (e.g. be dependent on seasonal, or wetting and drying conditions). The weights of the ensemble members determine the probability distribution of the hydrological simulation and derivatives such as the ensemble mean. In this research, both static and dynamic weighting have been tested.

The simplest method is by using equal weight. The merging by equal weight is calculated by:

$$Y_t^{eq} = \frac{1}{N}\sum Y_t^i \tag{3.1}$$

where Y_t^i is the output of ensemble member *i*, and *N* is the total number of model outputs.

Another method is to use past performance of each individual member to weight the member in the next period. The weight can be calculated by using normalized inverse distance to the perfect performance score, e.g. as follows:

$$W_1 = \frac{1/d_1}{1/d_1 + 1/d_2 + 1/d_3 + \ldots}, W_2 = \frac{1/d_2}{1/d_1 + 1/d_2 + 1/d_3 + \ldots}, \ldots \tag{3.2}$$

where W_1 is the weight for member number 1, d_1 is the Euclidean distance of the performance of the member number 1 to the perfect score, W_2 is the weight for member number 2, d_2 is the distance of the performance of the member number 2 to the perfect score, and so on. The sum of all weights, W_1+ W_2+ W_3+... is equal to 1. The Euclidean distance can also calculated for more than one dimension, with several performance metrics acting as the dimensions. The performance evaluation metrics are presented in the next section.

3.5 Performance assessment

3.5.1Single model performance

In order to calculate the single model performance, four scores are used. They are Nash-Sutcliffe efficiency (NSE) which gives the best model a NSE value of 1, with the equation as:

$$NSE = 1 - \left[\frac{\sum (Y_t^{obs} - Y_t^{sim})^2}{\sum (Y_t^{obs} - \overline{Y_t}^{obs})^2} \right] \tag{3.3}$$

percent bias (PBIAS) with the best value being 0:

$$PBIAS = \frac{\sum (Y_t^{obs} - Y_t^{sim})}{\sum (Y_t^{obs})} \tag{3.4}$$

RMSE-observations standard deviation ratio (RSR) with the best value is 0:

$$RSR = \frac{RMSE}{StDev_{obs}} = \left[\frac{\sqrt{\sum (Y_t^{obs} - Y_t^{sim})^2}}{\sqrt{\sum (Y_t^{obs} - \overline{Y}^{obs})^2}} \right] \quad (3.5)$$

Correlation coefficient which is calculated as:

$$r = \frac{\sum (Y_t^{obs} - \overline{Y}_t^{obs}) \sum (Y_t^{sim} - \overline{Y}_t^{sim})}{\sqrt{\sum (Y_t^{obs} - \overline{Y}^{obs})^2} \sqrt{\sum (Y_t^{sim} - \overline{Y}^{sim})^2}} \quad (3.6)$$

where Y^{obs} is the observed value and Y^{sim} is the simulated value; $\bar{Y}^{obs}$ is the mean of observed values; $\bar{Y}^{sim}$ is the mean of the simulated values.

To determine how well the model has performed, a NSE-based performance rating suggested by Moriasi and Arnold (2007) can be used. Although it was built for monthly time step, it will be used for other time steps as well.

Table 3-1. General performance ratings (Moriasi and Arnold 2007)

Performance rating	NSE	PBIAS (%)	RSR
Very good	0.75 < NSE < 1.00	PBIAS < ±10	0.00 < RSR < 0.50
Good	0.65 < NSE < 0.75	±10 < PBIAS < ±15	0.50 < RSR < 0.60
Satisfactory	0.50 < NSE < 0.65	±15 < PBIAS < ±25	0.60 < RSR < 0.70
Unsatisfactory	NSE < 0.50	PBIAS > ±25	RSR > 0.70

3.5.2 Ensemble simulation performance

Ensemble performance evaluation requires specific indicators. One of the most commonly used performance indicators is the Brier Score (BS), calculated by:

$$BS = \frac{1}{N} \sum (f - o)^2 \quad (3.7)$$

where f is the simulated probability of a binary event occurring that satisfies $Y_{sim} > Y_0$; and o is the observed occurrence of the event $Y_{obs} > Y_0$. The observed occurrences are scored as 1 if the condition of $Y_{obs} > Y_0$ is satisfied, and scored as 0 if not. The BS can be seen as a measure of error, similar to the Mean Squared Error for deterministic simulations, except that it is not measuring error in the simulated values but rather in the simulated probabilities for an event to occur. The lower the BS the better. (ECMWF 2015)

The BS can also be calculated for a single simulation result, e.g. the ensemble mean or any of the individual ensemble members, by putting either 0% (simulation is below the threshold) or 100% (simulation is above the threshold) as the simulated probability f. Note that this assumes that no uncertainty would be taken into account with the single value forecasts.

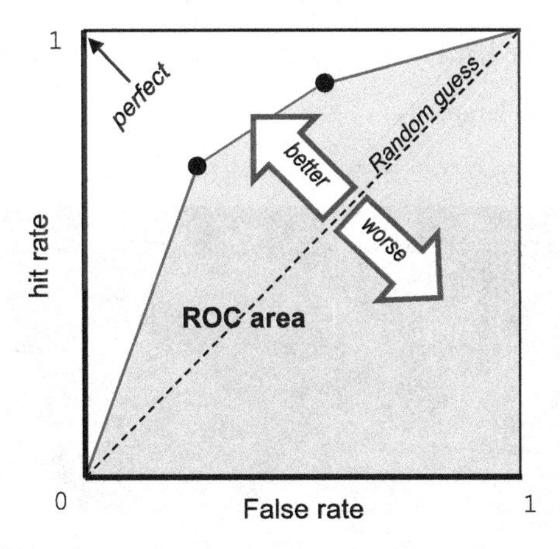

Figure 3-2. Relative operating characteristic (ROC) diagram, with ROC curve and ROC area

Relative operating characteristic (ROC) diagram (Figure 3-2) is used to analyse the probabilistic performance of the ensemble simulation in identifying an event threshold exceedance. The probabilistic property of the ensemble simulation is obtained from the normalized frequencies of the members that reported an event. The ROC diagram plots the hit rate against the false alarm rate (Georgakakos et al. 2004). The simulations lead to an alarm if the predefined threshold level is reached. If the alarm is proven to be correct when compared to measurements than it is a hit, but if the event did not occur, it is a false alarm. The hit rate and false alarm rate are calculated by:

$$HR = \frac{N_h}{N_e}, FAR = \frac{N_f}{N_e} \tag{3.8}$$

where the N_h is number of hits, N_f is the number of false alarms, and N_e is the number of observed events. The skill of a simulation is determined by calculating ROC area, the area under ROC curve for probabilistic simulation and the area under lines connecting lower left corner, model point, and upper right corner for deterministic simulation. A skilful simulation has ROC area more than 0.50 (ECMWF 2015).

3.6 Parallel computing as a facilitating technology

The proposed ensemble-based data-model integration framework may lead to extensive computing power requirements, especially with the use of a high resolution distributed model. In addition, a large amount of memory is necessary to temporarily store every hydrological state and variable for each grid cells for calculation of the next time step. Another requirement is the hard drive space for saving the selected hydrological state and variable which can be very large.

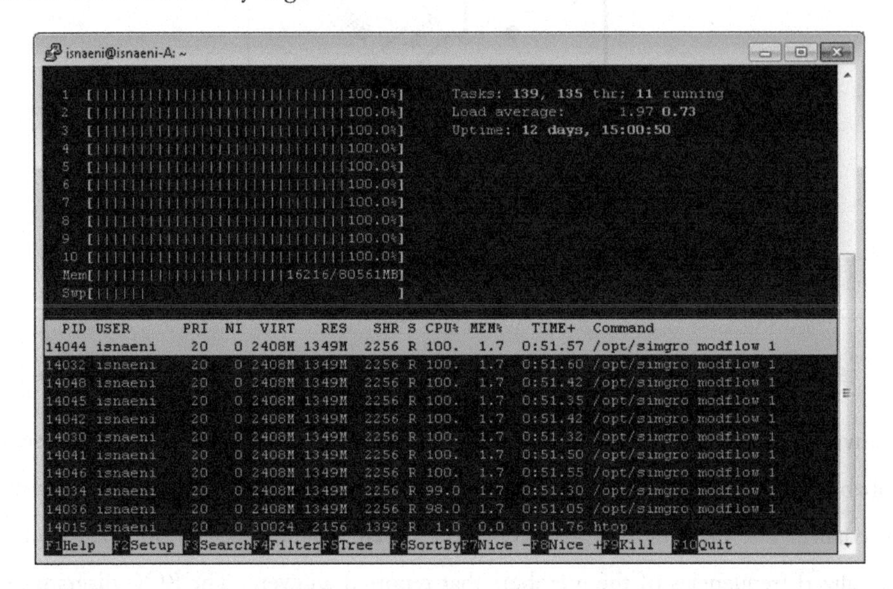

Figure 3-3. An example of the computational resource usage with a ten-core virtual machine

Since the model used in this study was developed for Intel-based PCs, the only solution to cope with this problem is to use parallel computing on multiple processors and/or PCs. In this research, a cloud computing server is employed to do the extensive computation and handling the high memory demand (Kurtz et al. 2017). The HPC cloud at the SurfSARA server (Surfsara 2014) was available to provide the computing power. The model can run in customable virtual machine (VM), where a number of CPU cores and memory can be requested. The parallelisation is handled by Python script on a Linux server. To control the VM, the Linux SSH (Secure Shell) communication protocol is used. For the MS Windows client, the SSH communication is done with the help of remote desktop software e.g. Putty and X2go. The file transfers are commenced by using

SCP (Secure Copy Protocol). Both SSH and SCP are secured communication protocols. All software that are used in the model runs are open source software that can be downloaded for free (Karssenberg 2010; van Walsum and Veldhuizen 2011; PuTTY 2017; Anaconda Inc. 2018; Canonical Ltd 2018; X2Go 2018).

Figure 3-3 shows a ten-core VM running the SIMGRO model, which has been running for twelve days. With a ten-core virtual computer, the computational time of a large number of model runs can be reduced by a factor of ten. The virtual computer can be instantiated several times to create several instances that can work together. The workload distribution for the VMs is handled with master-slave system. The system uses a master VM to control and distribute the work load to the slave VMs. The master VM also prepares the input files, collects the data, processes and stores the results accordingly.

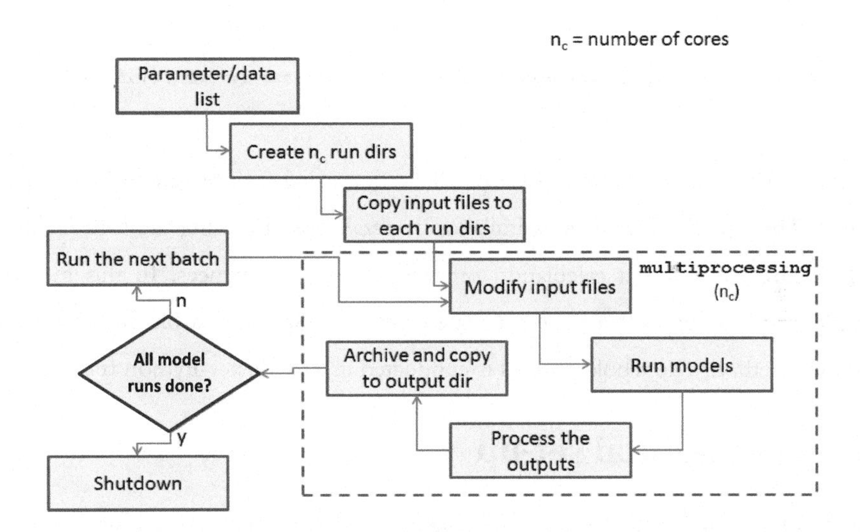

Figure 3-4. Parallelisation with multiprocessing module in Python

The flowchart of the calculation using multiprocessing Python module is presented in Figure 3-4. In this research the above method is utilised when computing several independent model runs. A run directory is created for each CPU core. The model input is prepared and copied to each directory based on the sampled parameters. Each of the CPU cores then computes a model in the run directory, when the model run is finished, the output files are compressed and saved in the output directory. Python modules such as Pandas, Matplotlib and Numpy are used in the process.

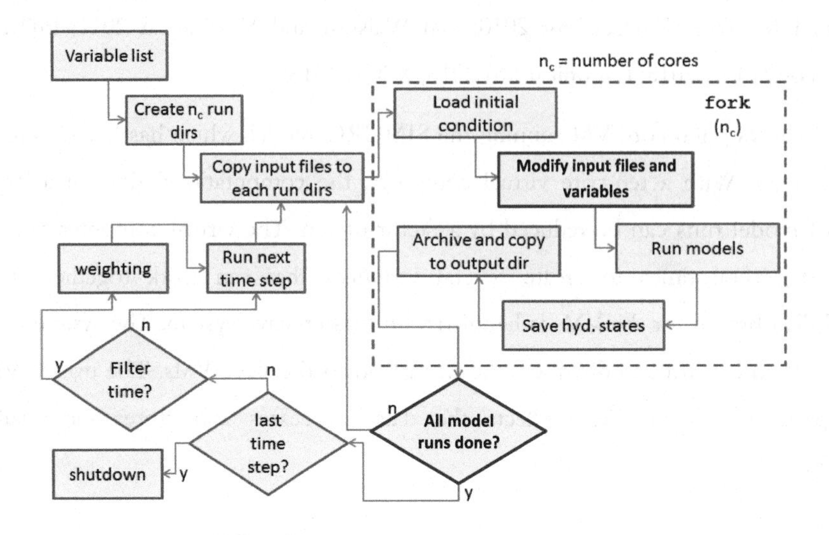

Figure 3-5. Parallelisation using fork feature in Linux; the fork command is called by the PCRaster-Python framework

Figure 3-5 presents the flowchart for parallelisation using fork feature in Linux operating system. The particle filter data assimilation is parallelised by using the fork feature. The fork feature will clone a calculation process to several instances. In this method the sampling is done by each of the calculated instances, which is a bit different from the previous method. The whole process is conducted in a PCRaster-Python framework.

3.7 Experimental set-up

The data-model integration framework (Figure 3-1) has been employed for the case study of Rijnland as described in the following chapters. The case study covers the following parts of the framework:

- Data selection and analysis
- Building and parameterisation of multiple models using different data sets for catchment characteristics
- Direct use of hydrometeorological data sources as model input (data-model integration method 1)

- Merging precipitation data sources and using merged data as model input (data-model integration method 2)
- EO data infilling and correction based on model output (data-model integration method 3)
- Data assimilation (data-model integration method 4)
- Creating a multi-model ensemble (24 members, based on 9 different data sources for catchment characteristics and hydrometeorological input)
- Performance assessment
- Static weighting for improving performance of the multi-model ensemble simulation
- Dynamic weighting for improving performance of the multi-model ensemble simulation

The EO information used consists of radar rainfall, land use, soil moisture, and actual evapotranspiration. The in-situ data include water level, ground water level, soil map, rain gauges, and pump operation. These data are presented in detail in Chapter 4.

For having a reference model, a so-called base model is built using the data sources normally used for this case study. The data sources for the base model are: precipitation from rain gauges, land-use from in-situ survey, generic soil maps from the study area, reference evapotranspiration from weather station, and other data required by the model, but only available from a single source, such as the aquifer characteristics. The base model is validated with discharge data, and additional validation is done using secondary data, such as ground water level and sub-basin discharge. The construction of the base model and validation results are presented in Chapter 5.

The development period for the base model is the full calendar year of 2010. The parameter values and ranges are obtained from expert judgement by the local water managers. The main validation, to observed total discharge from the catchment, is done for data of 2011, while the secondary validation is done using multiple data sources for the periods when the data is available.

The data-model integration methods experiments are presented in Chapter 6. Each experiment is done for the periods where the required data is available.

Finally, the ensemble simulation results, performance assessments, and the weighting for performance improvement, are presented and discussed in Chapter 7. These experiments are conducted for the period from October 2012 to October 2013.

Chapter 4. Case study and data sources

This chapter presents the Rijnland case study area and water system, and analyses the hydrometeorological data sources for Rijnland used in this research. The data sources consist of in-situ measurements, earth observation and remote sensing (satellite and radar), and model results. The data constitute land-use, soil maps, precipitation, evaporation, soil moisture, water level, and discharge.

4.1 Rijnland

4.1.1Catchment characteristics

The Rijnland area, located in the western part of the Netherlands, is a low lying area with most of it being reclamation lands from lakes and the sea. More than 70% from 1000 km^2 of the area is a flat region with the ground elevation lower than mean sea level (Figure 4-1a). These low areas are divided into what are called "polders" in the Netherlands (Figure 4-1b), where the low area is surrounded by high ground or a dike to prevent water to flow in. On the other hand, the excess water inside a polder cannot be discharged by means of gravity, hence a pump is required to discharge the excess water.

The sub-polders are drained by a series of canals, which in most of the sub-polders are small ditches arranged in parallel. Water in small ditches is then collected to larger canals and at the outlet of the sub-polder it flows to the polder's main canals through a pump or weir. The large canals in the polder are connected to the main storage basin (Figure 4-1b) through a pump. Part of the Rijnland area, at the western boundary close to the dunes, is above sea level, and discharges to the main storage basin by gravity, controlled by weirs. Woerden and part of Amsterdam are neighbouring areas that discharge excess water to Rijnland. The excess water in the main storage basin of Rijnland is discharged to the North Sea and rivers by four large pumping stations, namely: Gouda, Halfweg, Katwijk and Spaarndam as seen in Figure 4-1b. The total capacity of all pumps is 194 m^3/s (Rijnland Water Board 2014).

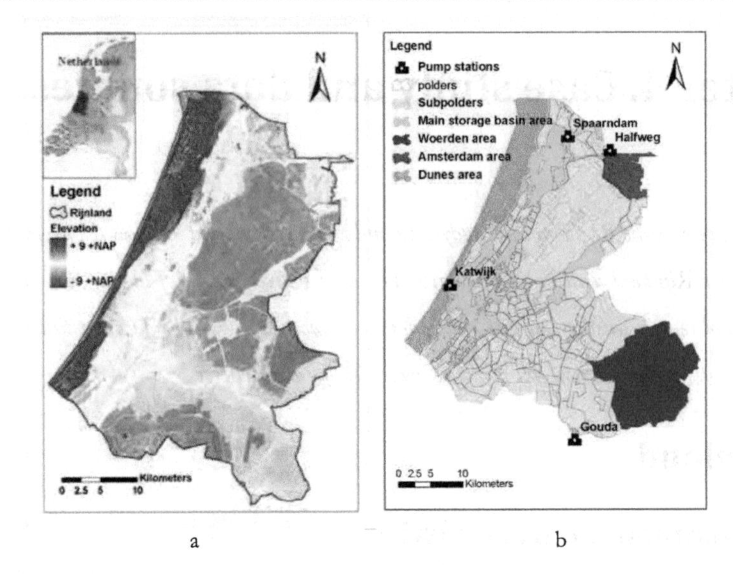

Figure 4-1. Rijnland elevation map, blue coloured is below mean sea level (a), and Rijnland polder and sub-polder map (b)

Rijnland Water Board (Hoogheemraadschap van Rijnland), the organization which is responsible for water management in the area, applies strict rules to the water-level control in the canals. Target levels are used to control the water level in every polder, with a specific value for each polder depending on the elevation and land use. The target water level is different in summer and winter, with the summer target level being higher than the winter level. This is necessary in order to have nearly constant ground water level in the middle of the field, which is very shallow in the lowland area. The automatic pumps will make sure that the water level in the canals is close to the correspondent target level. With this management, the water flows differently in wet periods compared to dry periods. In wet periods, the area has much excess water that needs to be pumped out, hence the water is flowing out from the canals to the rivers and sea. While in dry periods, the area needs water, hence the water from outside (from river or lake) is flowing to the tertiary canals. The two distinct water flows are presented in Figure 4-2.

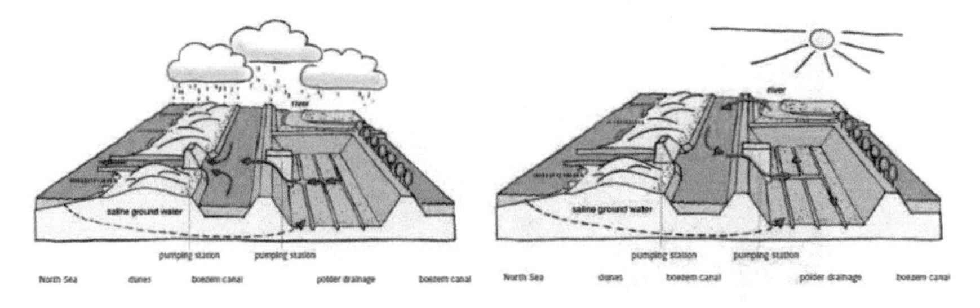

Figure 4-2. Water management scheme in the Netherlands (Source: Rijnland Water Board)

The lowland area is mostly several meters below mean sea level, while the target water level of the main storage basin is only -0.65±0.05 m NAP. These conditions lead to a high water pressure from the sea and main storage basin causing ground water flow toward the lowland, as seen in Figure 4-2. Figure 4-3 shows one of the areas in Rijnland, where the water level in the canal is higher than the field.

In Rijnland, the daily surface water balance (excess or deficit) is one of important information for operational water management by the Water Board. In wet periods where the system has high excess water, the managers need to know how much water they should pump out to avoid flooding. In dry periods where the water system needs water from outside the system, usually from river or neighbouring Water Board's area, the Water Board needs to know how much water that they need to let into the system. The water comes from outside also affecting the water quality in Rijnland which is another major issue for the managers.

Rijnland is a highly controlled water system. A controlled water system is a less natural system where flows and system states are regulated by control structures (van Andel et al. 2010). Inside the controlled water system, human influences highly affect the hydrological processes. For example, during a heavy rain, the water is pumped out from the system so a flood, which would have naturally occurred can be avoided. In Rijnland, the human influences go beyond that, such as pumping before the event, determined by forecasted heavy rainfall, or flushing of the system to maintain a good water quality. These human influences should be taken into account when interpreting monitoring data.

Figure 4-3. Example of canals in Rijnland, where the water level is higher than the surrounding fields

4.1.2Hydrometeorological data

Rijnland has many hydrometeorological data sets available. In general, these can be classified based on the data source, i.e. in-situ measurements and earth observation (satellite and radar). Figure 4-4 shows the data sets available and which part of the hydrological cycle they concern. Table 4-1 lists the metadata of the data sets used in this research. The data have different spatial and temporal resolution that might need to be altered in order to create an integration. The temporal and spatial coverage also differs, the satellite based data is available for a year starting at September 2012. The in-situ ground water level is available for the year 2010 and back. The other data has a long coverage up to 50 years back. The following sections describe each data set used in more detail.

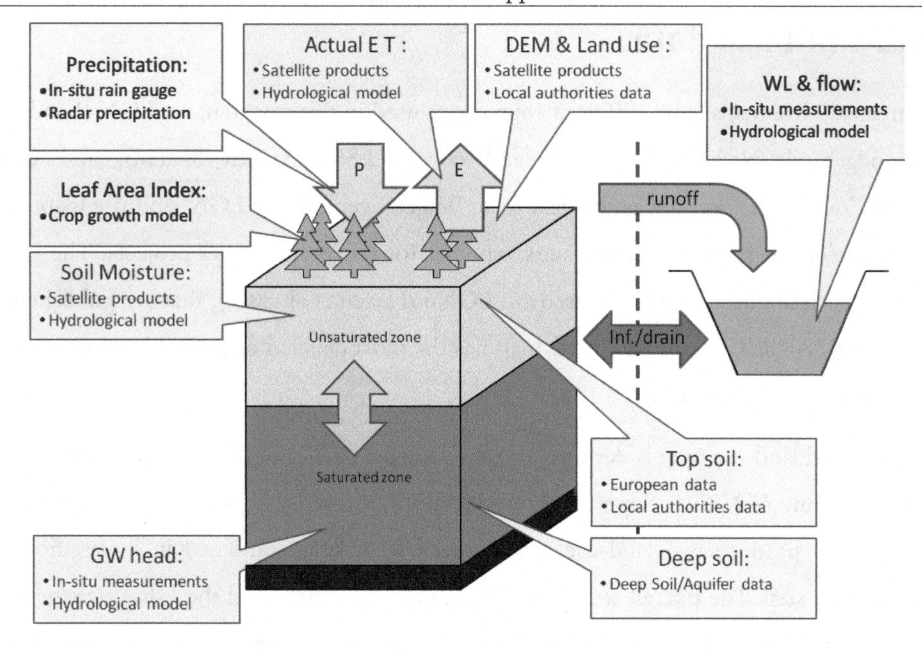

Figure 4-4. Available data and data sources in Rijnland. Not all available data sources were used in this research

Table 4-1. Available data sources in Rijnland that were used in this research

Name	Source	Temporal resolution	Spatial resolution	Unit	Period
Precipitation	In-situ rain gauge(KNMI)	Daily	/	mm/day	1950 to 2014
	Radar	Daily	1x1 km	mm/day	2009 to 2013
Actual Evapotranspiration	Satellite products	8-Day	250x250 m	mm/day	10/2012 to 10/2013
	Hydrological model	Daily	50x50 m	mm/day	2009 to 2013
Soil Moisture	Satellite products	8-Day	250x250 m	cm^3/cm^3	10/2012 to 10/2013
	Hydrological model	Daily	50x50 m	cm^3/cm^3	2009 to 2013
Discharge	In-situ measurements	10 minutes	/	m^3/s	1985-2014
	Hydrological model	Daily	/	m^3/day	2009 to 2013
Surface water level	In-situ measurements	10 minutes	/	m+NAP	1998-2014
	Hydrological model	hourly	/	m+NAP	2009 to 2013
Ground water level	In-situ measurements	Daily	/	m+NAP	2000-2012
	Hydrological model	Daily	50x50 m	m+NAP	2009 to 2013

4.2 Land use data

Three land-use maps from different sources are used in this research, i.e. LGN (Landelijk Grondgebruik Nederland-National land cover), LS5 (Landsat 5 TM), and GBC (GlobCover). The first one is the data from WageningenUR, the LGN land use map. The map is widely used in the Netherlands, whether for research or other projects. The LGN maps which are used in this research are LGN6 (Hazeu et al. 2010), that are published in year 2007/2008. The LGN land-use map has the most detailed maps and highest number of land-use classifications.

The second land-use map is derived from two Landsat 5 TM satellite images, acquired on 4/7/2010 and 6/9/2010. Supervised classification was used with a set of in-situ surveyed samples to produce two land-use maps, which were later combined in a classification refinement step. The overall accuracy of the final map is 88% and the k-hat statistic 0.83 (Nunes et al. 2013). The product has 30 meter resolution, which is further resampled into 50 meter resolution for the hydrological model (Chapter 5). The third land-use map is the GlobCover 2009 derived from ENVISAT MERIS satellite images, which is a global product at 300m spatial resolution (Arino et al. 2008). The three land use maps are presented in Figure 4-5, while the distribution of the land-use classification can be seen in Figure 4-6.

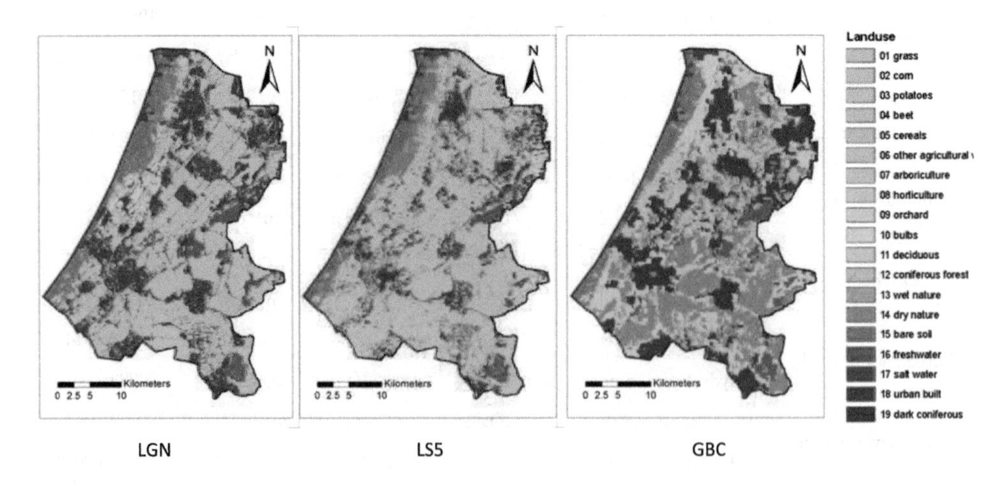

Figure 4-5. Three land use maps used in this study: LGN (Landelijk Grondgebruik Nederland-National land cover), LS5 (Landsat 5 TM), and GBC (GlobCover)

In Figure 4-5, the LGN map shows similarity to the LS5 map, while the GBC map shows significant differences. Looking at the comparison of LGN and LS5, difference can be seen in the western part near the dunes. The LGN classifies the area as flower field but the LS5 classifies the area as grass land. The other area that has different land use also suffers from the same misclassification, i.e. where the LGN classified as agriculture area the LS5 saw it as grass land. For the GBC maps, apart from the different resolution, most differences can be seen where the area is classified as a dry nature area while LS5 and LGN classified this as grass land or agriculture area. All classification differences mentioned above are not expected to have significant impact on the hydrological model (Chapter 5) due to the similar properties between those classifications.

Figure 4-6 summarises the area distribution of the simplified land-use classes. The 19 land-use classes are grouped into seven. The aforementioned qualitative differences between the three land-use maps also show in this figure. The LS5 has the largest grassland area, while the GBC has the smallest grassland area. However, the GBC has the largest bare soil area, which includes the dry nature area. The GBC has the largest forest area. Looking at the urban area coverage, the LGN and GBC have a similar area coverage, the LGN has a smaller urban area. Water bodies coverage is higher in the LGN map.

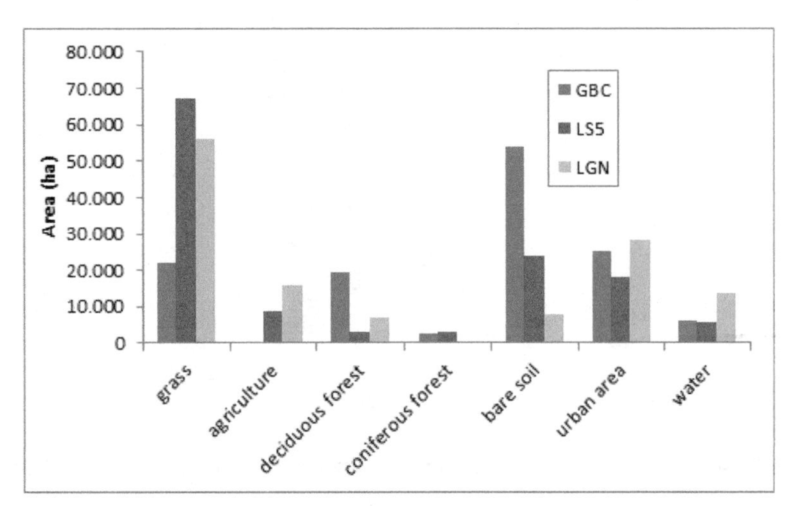

Figure 4-6. Distribution of land-use classification of the three land use maps. The classification is simplified into 7 classes.

4.3 Soil data

The soil map for Rijnland is available from two sources: European database (EDB) and Dutch database (DDB). The EDB map is developed from the Soil Geographical Database of Europe (EC 2003). The database is transformed into spatial maps of soil hydraulic properties using new pedotransfer functions (Tóth et al. 2015). The pedotransfer functions are derived from the European Hydropedological Data Inventory (EU-HYDI) (Weynants et al. 2013). These maps are then transferred into soil properties for the hydrological model in this study (SIMGRO, Chapter 5) based on the similarity of the hydraulic properties.

The second soil map is obtained from the Dutch database (de Vries et al. 2003), and is widely used by the local authorities for research or projects. Because the soil classification of SIMGRO is the same as the Dutch database, there is no need of soil classification conversion. The two soil maps are presented in Figure 4-7. From the 21 classes of soil, the Rijnland area has 17 soil classes according to the DDB map, but only eight visible classes according to the EDB map.

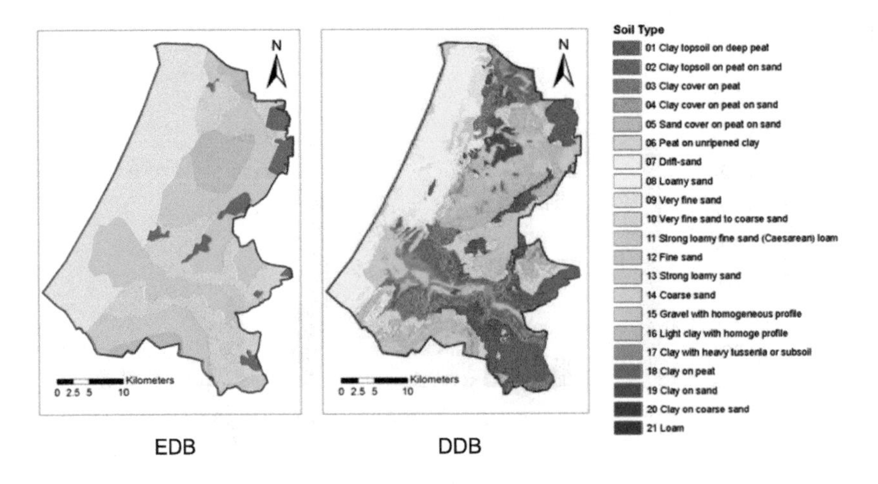

Figure 4-7. Two soil maps used in this study: from the European database (EDB), and from the Dutch database (DDB)

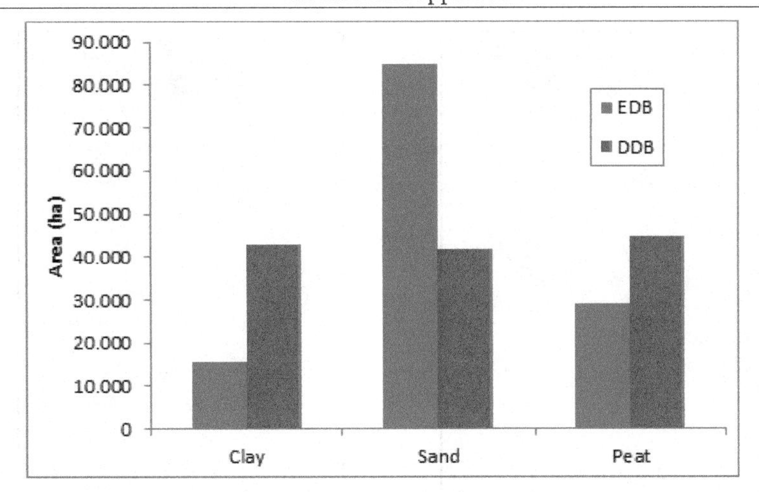

Figure 4-8. Distribution of soil type of the three soil map data. The classification is simplified into 3 classes.

Figure 4-8 presents the area distribution of the soil type where the 21 soil classes are grouped into three main classes, i.e. clay, peat and sandy soil. The DDB soil map has the Rijnland area almost equally divided into the three soil types. On the other hand, the EDB soil map has the sandy soil as the most common soil, and the clay as the least occurring soil type in the Rijnland area.

4.4 Precipitation data

The two precipitation data sets, gauges and radar, both consist of processed data, meaning there are no missing values and that they are calibrated to a reference data set. Hence, the two data sets are not expected to give very different values in recorded precipitation. Snow occurs only sporadically, such that the precipitation is mostly rainfall.

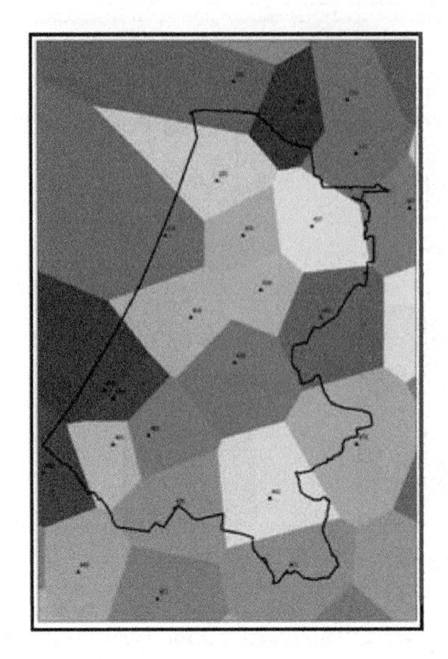

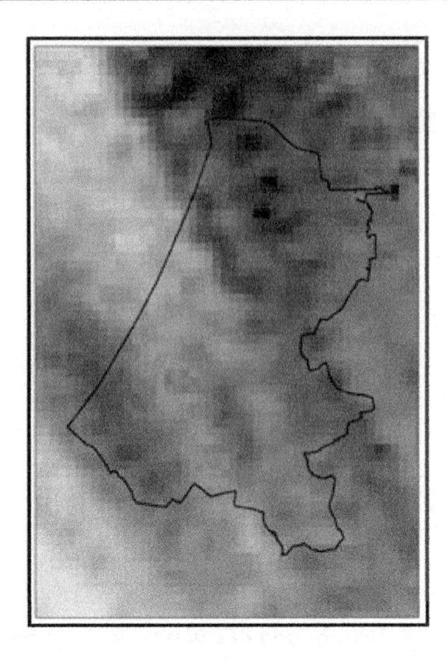

Figure 4-9. Rainfall maps, Thiessen polygon and radar grid cells

The gauge data are collected from the website of KNMI (Royal Dutch Meteorological Institutes, KNMI 2014). The 21 rain gauges are organized into a Thiessen polygon as shown in Figure 4-9. The radar data are obtained from Hydronet database (Hydronet 2014), with resolution of one kilometre, the grid cells over the Rijnland area can be seen in Figure 4-9.

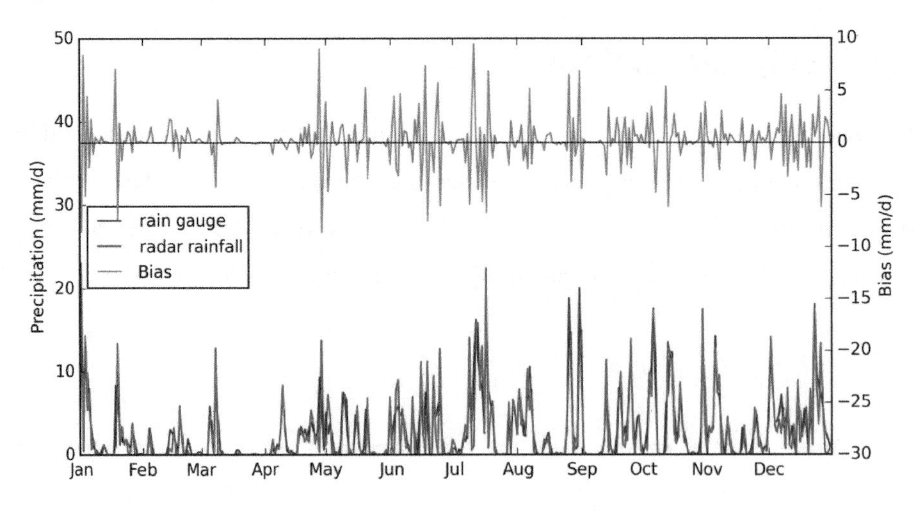

Figure 4-10. Comparison of gauge and radar area-average precipitation in Rijnland in 2012

Figure 4-10 presents the comparison of the area rainfall over Rijnland between rain gauge network in the form of Thiessen polygon and the radar rainfall data. The comparison shows that, although the pattern is the same, differences are observable, especially during high peaks. However, from the graph it can also be seen that the differences may often be a shift in time, where a positive difference is followed by a negative one the next day. This may be caused by the time difference in the daily precipitation data acquisition, i.e. radar data at midnight and the rain gauge in the morning, for which only a crude correction was made (moving 1/3 of the day's precipitation for radar data to the previous day).

Figure 4-11 presents the comparison of the spatial variability of monthly precipitation between gauge and radar in 2012. The comparison shows that the 1-km resolution radar is able to capture rainfall variability better than the rain gauges. For example in April 2012, there was much rainfall covering a small area in the central part of Rijnland, captured by the radar but not seen in the gauge network.

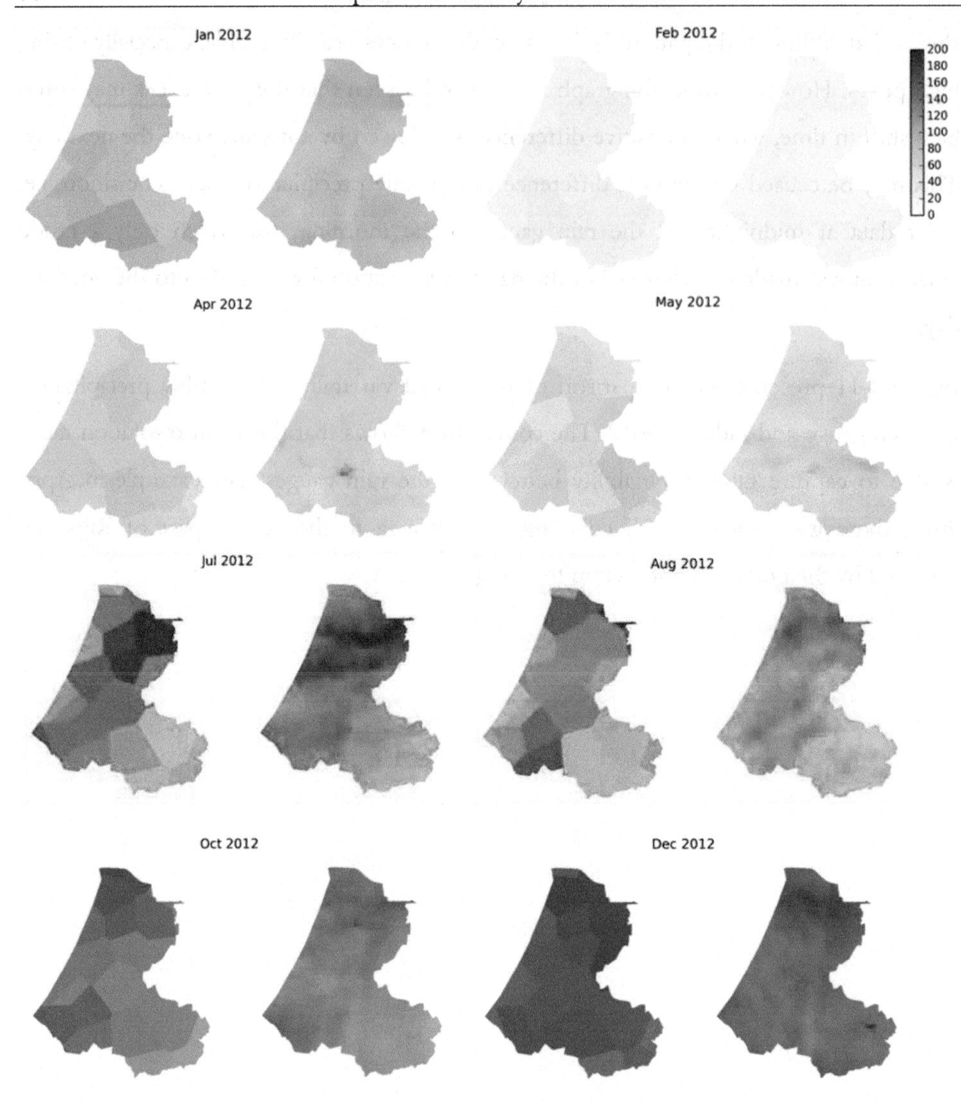

Figure 4-11. Spatial comparison between the monthly precipitation from Thiessen polygon and radar in mm

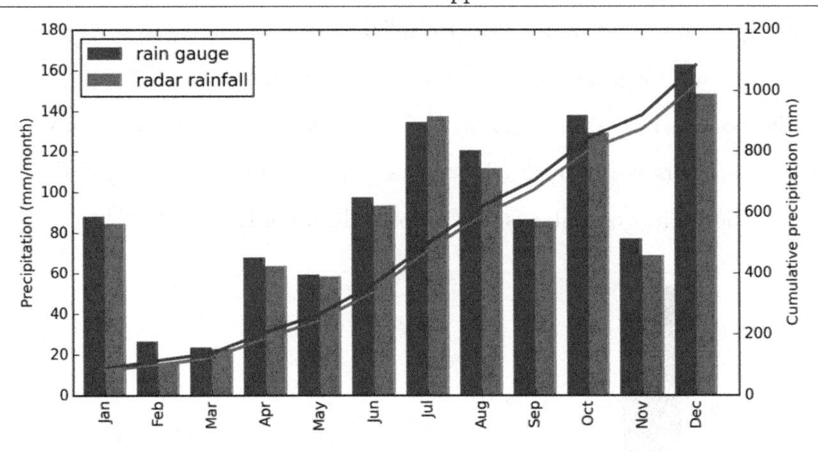

Figure 4-12. Monthly comparison between rain gauge network and radar area-average precipitation in Rijnland for 2012

Figure 4-12 shows the bar chart of the monthly area-average rainfall of the gauge network and radar for 2012, together with the cumulative time series. The rain gauge network estimates more rainfall in 2012 in almost every month compared to the radar data.

4.5 Evaporation

For this study, the Schiphol meteorological station is the main source of the reference evaporation for the Rijnland area. The reference evaporation is Makkink reference evaporation, widely used to calculate evapotranspiration in the Netherlands area (de Bruin and Lablans 1998).

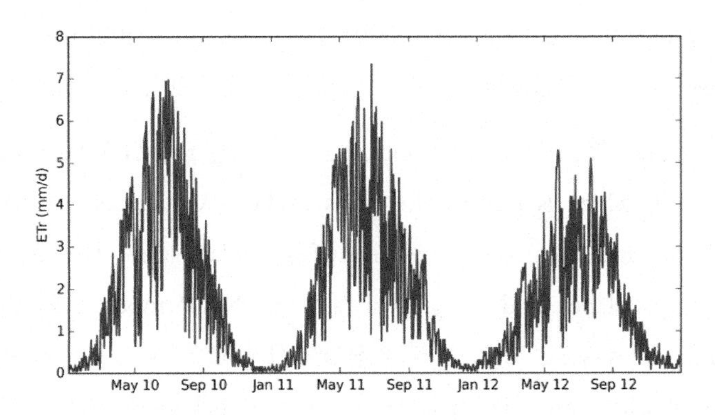

Figure 4-13. Reference evaporation in Schiphol meteorological station for 2010 to 2012

Figure 4-13 presents the daily ET_r data from Schiphol meteorological station for 2010 to 2012. The ET_r were high in the summer management period from April to September, and low in the winter period. The yearly ET_r was decreasing during the three year period. Figure 4-14 presents the normalised frequency distribution of the daily ET_r during the three year period. The graph shows that almost half of the ET_r was less than 1 mm/d.

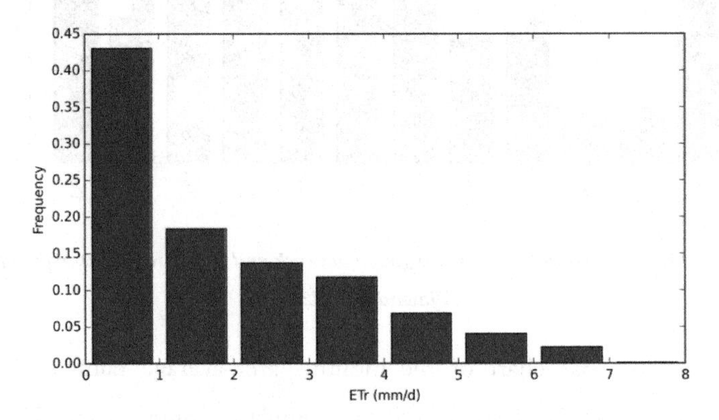

Figure 4-14. 2010 to 2012 ETr frequency distribution

The EO ET_a maps are derived from the Terra/MODIS satellite products by solving the surface energy balance equation (Cherif et al. 2015). Three Terra/MODIS products were required for the estimation of evapotranspiration: Surface Reflectance (MOD09Q1 at 250m, MOD09A1 at 500m resolution), Land Surface Temperature (LST), and Emissivity (MOD11A2 at 1 km resolution).

There were two kinds of EO ET_a maps available for the Rijnland area in this research: composite daily ET_a map and cumulative eight-day ET_a map. Both maps were collected from MyWater project (Araújo and Nunes 2012; MyWater project 2014). The ET_a maps have a 250 m resolution and cover the period from October 2012 to September 2013.

The composite daily ET_a maps were obtained by using the ITA-MyWater algorithm (Cherif et al. 2015), which is part of the adjusted ITA-Water tool (Alexandridis et al. 2009) for MyWater case study sites which includes Rijnland area. The method includes solving the energy balance and merging data from NWP models. In this case, the meteorological model was CPTEC/INPE (CPTEC/INPE 2014), a high resolution NWP model built for Rijnland area which is nested in a global model. The CPTEC/INPE model is an improved version of ETA model (Mesinger et al. 2012). The composite maps were accompanied by day of the year (DOY) maps, which specify which day the ET_a is for.

Eight-day cumulative ET_a maps were derived from the composite ET_a maps. A reliable method for temporal integration is used to obtain the missing daily values, as described by Alexandridis et al. (2009). The daily ETr for obtaining the missing values was calculated using standardised Penman-Monteith method (Allen et al. 1998). The evapotranspiration fraction (ET_rF), which is the fraction of ET_a and ET_r, was assumed as constant throughout the eight-day period. With this assumption, the missing data could be calculated as long as there is one ET_a value available in the grid-cell during the eight-day period.

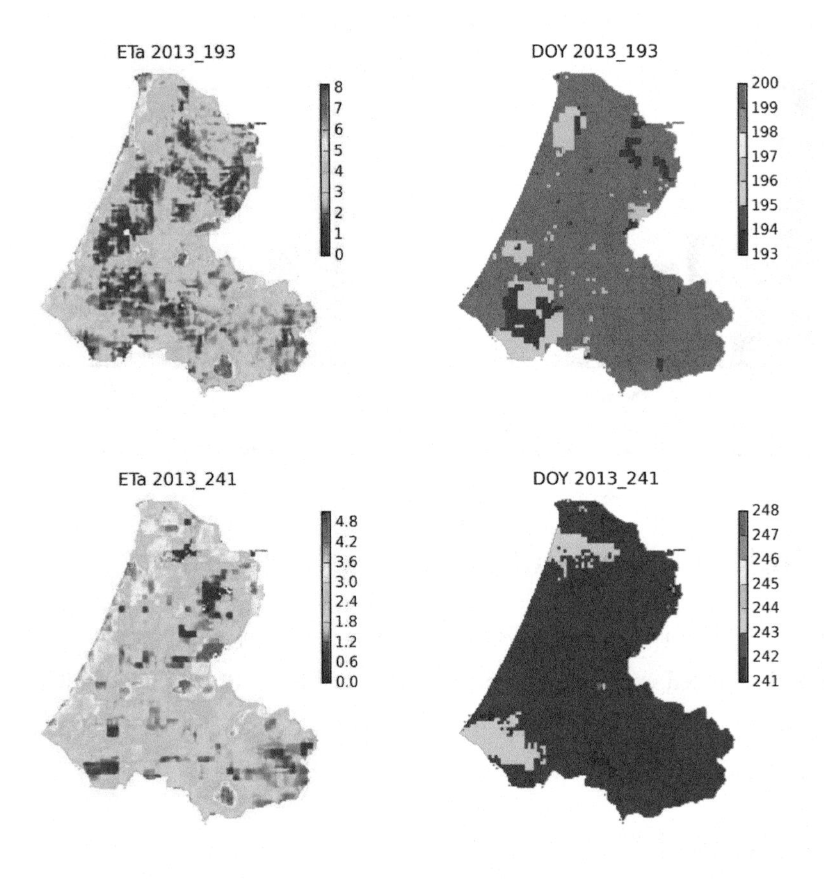

Figure 4-15. Examples of ET_a maps with corresponding DOY maps: 12 to 19 July (2013_193), and 29 August to 5 September (2013_241)

Figure 4-15 presents examples of the 45 EO ET_a maps from the MyWater implementation period, October 2012 to September 2013. Two ETa maps are presented, which have information for 12 to 19 July and 29 August to 5 September. Each map has its corresponding DOY map to inform which day each pixel has the ET_a information for.

The ET_a estimation is comparable to the ETr information from Schiphol for those dates. The large water bodies (lakes) were constantly giving the highest ET_a, as given by the reddish colours, compared to other areas. Figure 4-16 shows the normalised frequency distribution of the EO ETa pixels for the whole implementation period. Both daily composite maps and eight-day cumulative maps are presented. The frequency distribution has a pattern similar to the pattern of ETr. This shows that the EO generally provided reasonable estimations of ETa in the Rijnland area.

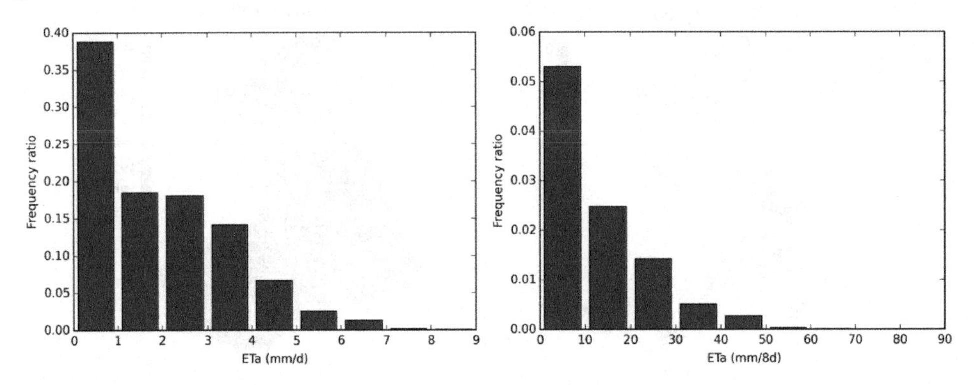

Figure 4-16. Normalised frequency distribution of daily EO ETa (composite maps) and the eight-day EO ETa

4.6 Soil moisture and ground water data

Root zone soil moisture data was estimated from EO by using the evaporative fraction maps and saturated water content map, as described by Scott et al. (2003). The EO data source is the same as for the evapotranspiration map. The root zone soil moisture was calculated by:

$$\theta_{rz} = \theta_{sat} * \exp(\Lambda - 1.0) / 0.42) \quad (4.1)$$

where θ_{rz} is root zone soil moisture, θ_{sat} is the saturated water content and Λ is the evaporative fraction. Similar to the ET_a maps described in the previous section, the soil moisture maps are composite maps, produced every eight-day period. The maps have the same DOY maps as the ETa maps.

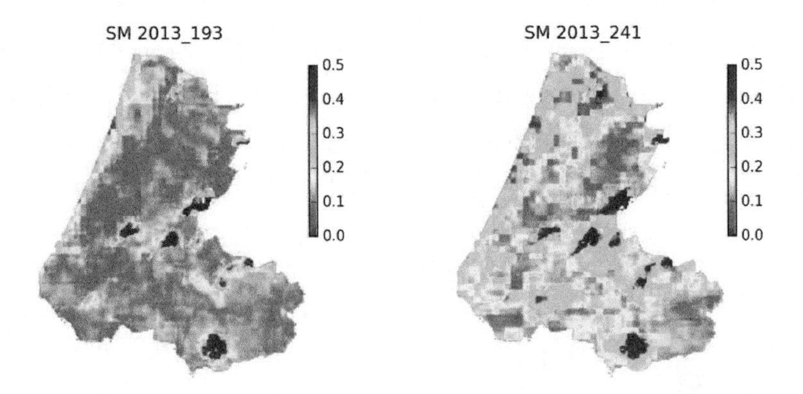

Figure 4-17. Examples of root zone soil moisture maps in m^3/m^3: 12 to 19 July (2013_193), 29 August to 5 September (2013_241)

Figure 4-17 presents two of the 45 EO root zone soil moisture maps. The maps seem reasonable with the water bodies having high soil moisture content. However, the frequency distribution presented in Figure 4-18 indicates that most of the Rijnland area would have a low soil moisture content, below 0.20, most of the time, which does not match with the shallow ground water level characteristics of the catchment where the soil should be often saturated.

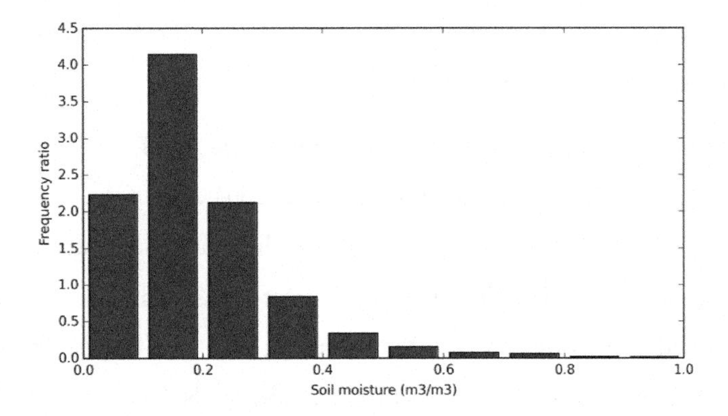

Figure 4-18. Normalized frequency distribution of eight-daily EO ETa

Rijnland area has many ground water observation wells, however, most of them are located near a canal which influences the shallow ground water head. The observation data were obtained from DINOLoket (2013). Figure 4-20 shows the locations of the measurement wells that have data between 2010 and 2013. The locations are mostly concentrated in the northern dunes and Woerden area. Due to the nature of the Rijnland water system, the ground water level can abruptly change over a short distance, especially between areas with high differences in target water level for the canals.

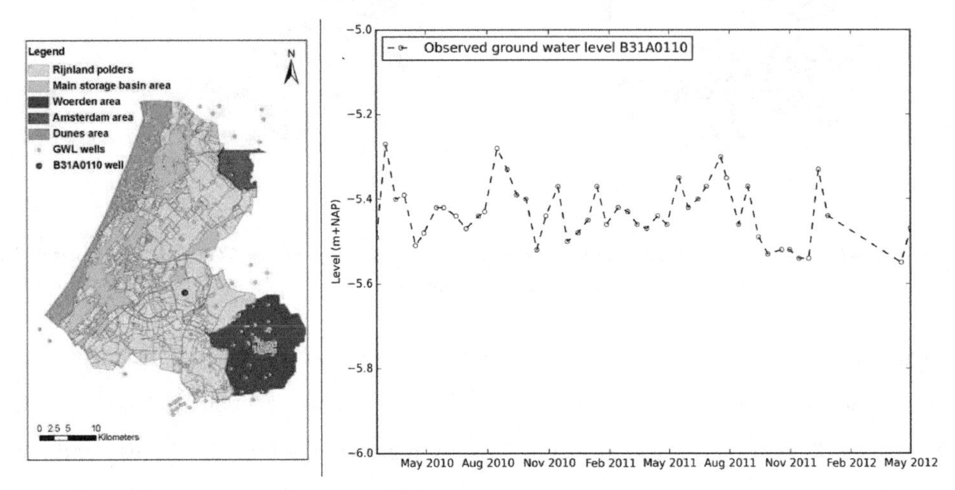

Figure 4-19. Ground water monitoring wells in Rijnland that have data from 2010 onward (left) and data from the observation well B31A0110, located in the centre of Rijnland (right)

Figure 4-19 also presents one of the ground water level time series in the central part of Rijnland, for well ID B31A0110. The variability is around 30 cm, with no recognisable seasonal pattern during two and a half years.

Aside from the point-based ground water observation wells, the Dutch also have ground water reference maps called GXG (De Gruijter et al. 2004). The GXG map-set consists of three maps: GHG with the average of the three highest ground water levels of the year, GLG with the average three lowest ground water levels, and GVG with the average of the three ground water levels on 14th, 28th of March, and 14th of April. The GXG map-set is widely used in ground water analysis in The Netherlands. In this research, the GXG map-set from NHI (National Hydrological Instrument) model results was used (NHI 2014) for comparison with the Rijnland SIMGRO ground water model results (Chapter 5). The map-set can be seen in Figure 4-20. The NHI model is recognised as a good hydrological model that simulates ground water level for the whole of the Netherlands. The resolution of the model is 250 m.

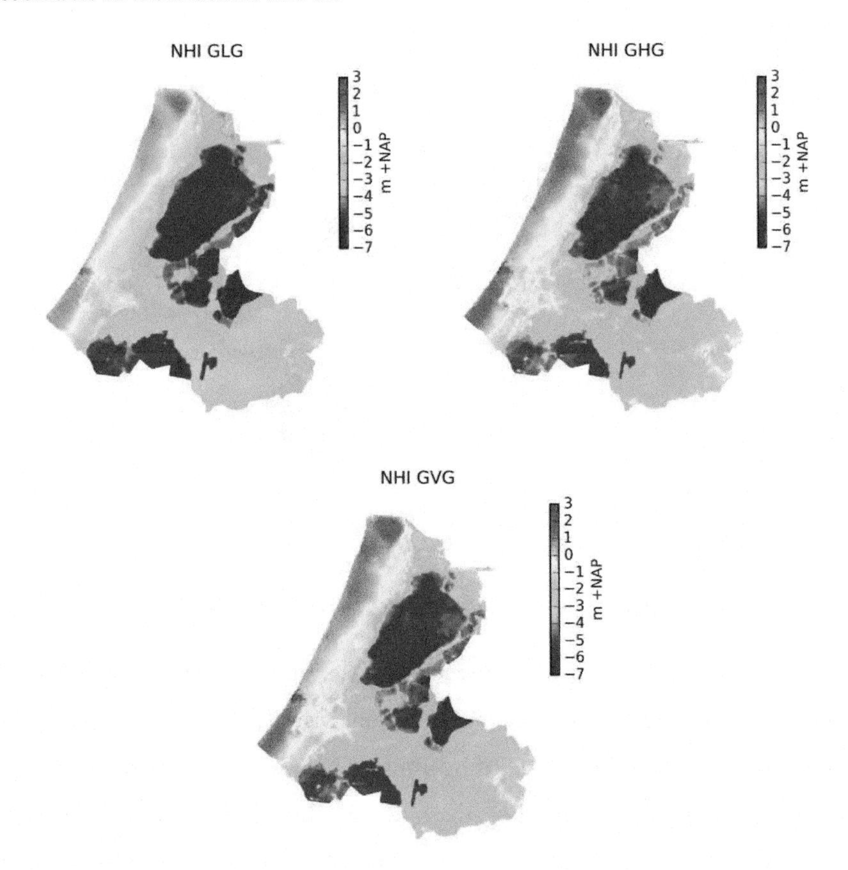

Figure 4-20. GXG map-set , GLG (low ground water level), GHG (high ground water level) and GVG (summer ground water level) from the NHI model (De Gruijter et al. 2004; NHI 2014)

Figure 4-20 shows that, overall, ground water levels are high in the dunes area near the North Sea where the terrain elevation is high. The lowest ground water levels appear in the low-lying reclamation areas in the northeast and southern parts of Rijnland.

4.7 Field survey for EO data calibration in Rijnland

In order to calibrate and validate the EO data, a field survey was done. The survey was done in three periods: 17 to 19 July 2012, 26 to 28 June 2013, and 23 to 26 September 2013. The first period was for calibration, while the second and third period was for validation in summer and winter season. The surveys were conducted in 28 locations, selected based on land-use, soil type and accessibility. The 28 locations can be seen in Figure 4-21.

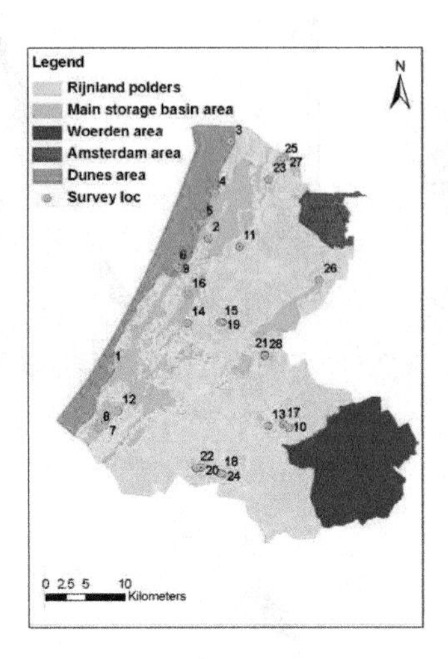

Figure 4-21. Field survey locations for EO data calibration in Rijnland area

The field survey comprised of three observations, i.e. land-use, leaf area index (LAI), and soil moisture. The land-use observations were done visually, by observing the land cover. The LAI observations were conducted by using a fish-eye camera to capture 360^0 images of leaves and grass, which were then analysed with dedicated software. The images for high trees were taken by upward looking camera, while for grass and shrubs, downward looking pictures were taken. The soil moisture measurements were conducted by using a TDR 300 soil moisture meter (http://www.specmeters.com/soil-and-water/soil-moisture/fieldscout-tdr-meters/). The soil moisture estimation analysis can be seen in Alexandridis et al. (2016).

The field survey of land-use indicated that the most observed crop is pastures for grazing. The natural vegetation constitutes small patches of broadleaves and few coniferous. The dunes are covered by natural herbaceous vegetation with open sand areas. There are many urban areas with lots of green space. The LAI field survey found that the survey locations have a high degree of vegetation coverage, fully covering the ground even in summer. The full coverage of grass made the estimation of LAI more challenging, however, for the trees the condition in Rijnland was favourable, with low sunshine giving high contrast to the upward looking vegetation images. The soil moisture survey faced difficulties due to the shallow ground water level and ponding water was found in most of the locations.

4.8 Summary

The data analysis showed that all data sources have estimated reasonable values for the hydrometeorological variables. The precipitation estimated by the radar rainfall has similar patterns to the one from the rain gauges, although some differences are observable. The EO estimation of ETa was in the normal range typically observed in the Netherlands. The different sources of land-use map gave visible differences, with some incorrect land-use classification e.g. grass with potatoes, grass land with dry nature area. However, these kinds of error are considered acceptable, since land-use characteristics are still similar to each other. Other differences occur on the different soil map data sources. In general, for the data sets evaluated, it cannot be judged that one is better than the others.

Chapter 5. Model development

This chapter presents the model development for the case study. The Rijnland SIMGRO model was validated against the main output variable, which is the discharge at the main outlets. However, there were several other observational data-sets available for the case study area, hence the model results were further validated against these additional data, i.e. individual polder's discharge, surface water level, ground water level, evapotranspiration, and soil moisture. Confidence in accuracy of some of these data sets is low, therefore, a mutual validation is done between measurements and model result. Uncertainty analysis results are presented.

5.1 Introduction

A base model (reference model), using the data sources normally used for the case study, is developed. The objective is to simulate the main storage basin discharge from the Rijnland water system. The SIMGRO model uses parameter values suggested by experts from the Rijnland Water Board. A limited manual calibration was done to choose parameter values when a range was provided by the experts. Apart from the main storage basin's discharge data, there are several other data sets available to further validate the model. Discharge time series of some of the local polders were available, although derived from the pump activities instead of directly measured data. Actual evapotranspiration maps from EO data are compared with the modelled evapotranspiration for point locations and for the whole catchment. Soil moisture maps from EO data are compared with the SIMGRO results in a similar way. Lastly, ground water level measurements and national-scale ground water model results, are compared on a point-by-point and catchment area basis.

These data sets were compared to the simulated results as a secondary validation of the model. However, these data sets have a degree of uncertainty; therefore, they cannot be held as a ground truth to solely evaluate the model. Especially the EO data may have high uncertainty, being indirect measurements and, in this case study, being affected by cloud cover that can be severe over Rijnland. As neither the validation data, nor the model results can be considered as ground truth, the validation goes both ways. In this thesis, this two-way validation is called mutual validation.

In the following sections the model development steps mentioned above are described in detail.

5.2 SIMGRO model set-up

In this research, we built the Rijnland SIMGRO model on a 50 meter resolution, with 125 km^2 area forming more than 500,000 active grid cells. The selection of the 50 meter resolution was due to the characteristics of the drainage canals that generally are built in parallel every 30-100 meter. The active cells cover the whole Rijnland area and additional areas of Woerden and part of Amsterdam, which also discharge to Rijnland.

The surface water system comprises a complex of drainage and discharge canals, clustered in sub-polders. The model has more than 700 sub-polders, which are grouped into around 200 polders. The water system control rules were derived from the Rijnland water board database. The control information covers target level for each polder, pumping capacity, pumps operation, weir operation and the layout, dimensions, and connections of the canals. However, in several locations, the input information was not available and needed to be assumed, such as information on the small local weir capacity and operation.

A DEM derived from AHN (Actueel Hoogtebestand Nederland) Lidar data (1 meter resolution) was used to define the ground surface elevation of the model (AHN 2011). The Lidar data was re-sampled to 50 meter resolution by simple averaging. LGN 6 land-use map (Hazeu et al. 2010) was chosen as the land-use map for the base model. The soil map uses EDB map (Tóth et al. 2015) as discussed in Chapter 4.

The land use data was converted to land use index as required by the METASWAP part of SIMGRO. Some of the land use classes were simplified into available SIMGRO's land use classes. For example, various forest types were grouped as one deciduous forest. The road, ports, and cities were grouped into urban build class, assuming 40% of the area was unpaved. Hence, 1000 m^2 area out of 2500 m^2 of each urban cell behaves as a normal grassland cell, with normal hydrological processes occurring, such as rainfall water to infiltrate, ponding water evaporates, etc. The remaining 60% of the urban area was assumed to be impervious, draining straight to the sewer system, treatment plant, and main storage basin. The soil map was also converted to meet the standard soil classification used by SIMGRO. Soil types with similar properties were grouped into one class according to the available conversion table provided by the Dutch authorities.

The deep soil data for MODFLOW was obtained from the NHI (National Hydrological Instrument) data, which is a model build for the whole country of the Netherlands (NHI 2014). The MODFLOW model has seven aquifer layers with the spatial data of 250 meter resolution. The ground water boundary is located along the edge of Rijnland area and acts as model boundary. The boundary is situated along the border with other water boards and the North Sea. The boundary of the MODFLOW part of the model was assumed to have fixed ground water head. The pressure head was determined by average ground water head obtained from NHI data. The model boundary can be seen in Figure 5-1 along the thick black line.

The inflow and outflow from/to the model domain was determined by the target water level in the modelled area. When the water level in a surface water area is lower than a given target level, a supply from outside of the domain to the polders is triggered. The outflow and inflow are limited by pumping and inlet capacity respectively.

The target water level is higher in the summer than in the winter. At the beginning of summer or winter, the target water levels in the model are changed at the same time for all polders, which does not occur in reality. The farmers are responsible to open the local inflow gate when the water level is set-up, and the authorities do not have information on when or how much the inflow is. However, when the volume of inflow water is too high and the water level exceeds the target level, the water will be pumped back to the main storage basin.

Each sub-polder is treated as one surface water storage, meaning it has one surface water level for all cells located in that sub-polder. The main storage basin is also modelled as one big surface water storage, although it is spatially spread across the area; it has the same water level everywhere, neglecting the wind pressure and routing time. This is acceptable because the model time-step is daily and because Rijnland Water Board's control targets are based on area-average level.

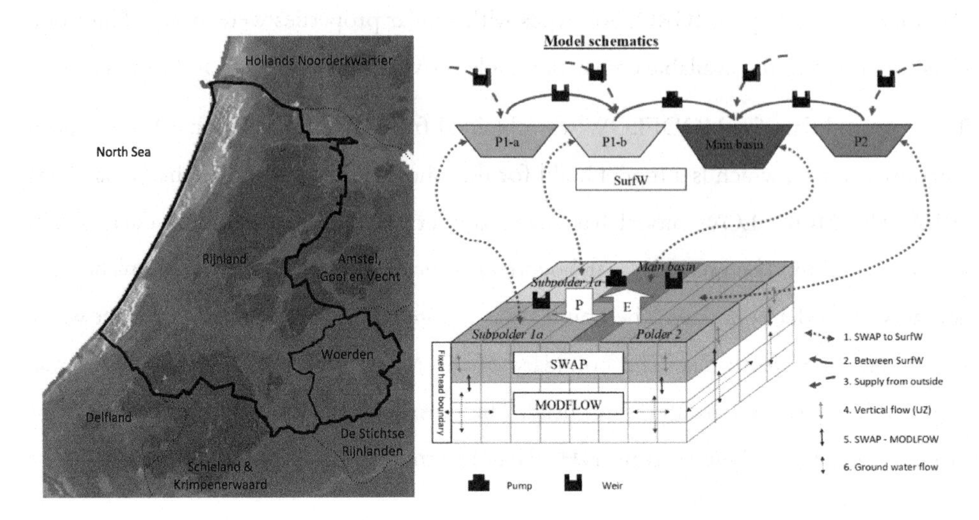

Figure 5-1. Rijnland model boundary and schematization

Each surface water system is connected through pumps or weir, through which they exchange water with each other. Most water systems pump to the main storage basin, except the dunes area, where the water flows from the model domain directly out, and the adjacent western parts with surface levels above the main storage basin level, where the water flows directly to the storage basin.

The pump operation in the model relies on the water level in the surface water unit. It will start pumping when the water level is higher than target water level in the corresponding surface water system. The pumping capacity is set according to the pumping station data from Rijnland Water Board.

The capacity of inflow weirs to sub-polders was unknown, hence it was assumed to have 0.2 m^3/minute for each hectare of polder area downstream of those weirs. In reality, the inflow could come from several weirs, but in this model, it had been simplified to have only one weir for each sub-polder. The weirs also operate based on difference between the target water level and calculated actual water level in the sub-polder.

The main storage basin has four large pumping stations: Gouda, Halfweg, Katwijk, and Spaarndam pumping station. Nevertheless, the model was developed to use only one pumping outlet with a capacity of the total capacity of those four pumps (point outflow boundary). The main pump operates with the same principle as sub-polder pumps: it would discharge water if the water level is higher than targeted level. The same assumption applies to the inlet for main storage basin. Furthermore, it was assumed that an unlimited water supply is supplying the main storage basin for any lack of water. Flushing of the main storage basin was not modelled, assuming water that is pumped in for flushing on one side (Gouda pumping station) would at the same time be pumped out at the other side. The observed inflow through Gouda station is subtracted from the total observed discharge at the other three pumping stations, such that also from the observed discharge time series flushing of the main storage basin is removed.

The total discharge of the four main pumping stations was chosen as the main output variable for model simulation. Using the total discharge, unrecorded human influence such as gate opening for inlet to the polders was considered minimum as these mostly occur at the sub-polder level. However, the water is originating from the main storage basin, and the excess water in the polders would be pumped back to the storage basin. Therefore, the unrecorded manual control does have an effect on the mid-term (daily to three-day) water balance of the main storage basin.

The precipitation input was obtained from the rain gauges network of KNMI, with 21 stations supplying the data as explained in Chapter 4. The rain gauges were transformed to spatial data with Thiessen polygon method. The transformation was done by assigning grid cells to the nearest rain gauge location. Reference evapotranspiration (*ETr*) time series was obtained from the Schiphol Weather Station. Actual evapotranspiration was calculated by the model using the Makkink method.

The model was run with parameter values suggested by modelling experts of the Rijnland Water Board such as for the inflow and outflow resistance. A limited manual calibration was done when a range of suitable parameter values was provided.

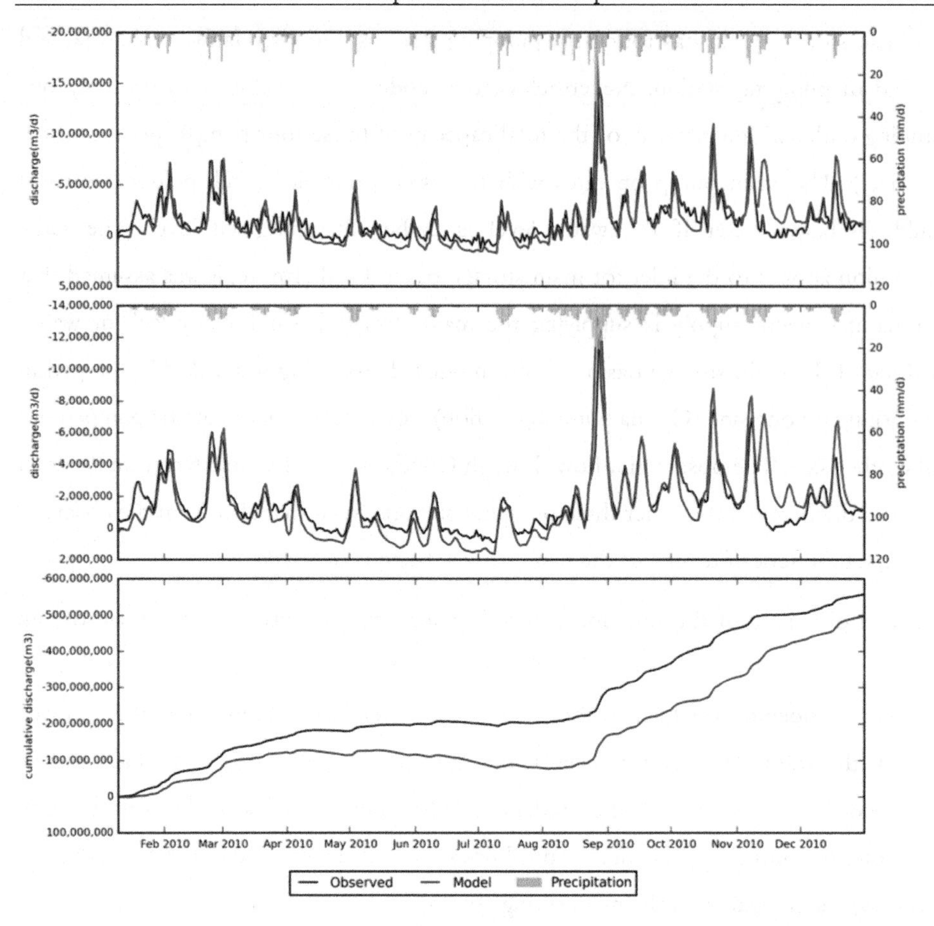

Figure 5-2. Discharge simulation results compared to observed data

The simulated daily discharge is compared with measured discharge in Figure 5-2. The first panel shows the daily discharge, middle panel presents three-day moving-average discharge time series and the last panel - the cumulative discharge. Using moving average allows for filtering out the human influences. The figure shows that the model overestimates positive discharge, meaning overestimation of water that flows into the catchments. The overestimation can be seen both in the daily and cumulative graph. The daily graph shows several deviations in the dry period between April and August, while the cumulative graph has the modelled line shifting away from the observed one in the same period. However, the model simulates the negative discharge more accurately, especially the peaks when high precipitation occurs in the catchment. Even in the dry periods, when a precipitation event occurs, the peak is correctly simulated by the model, such as the peak in May. In general, the discharge is under-estimated in summer and over-estimated in winter.

The effect of human influence (operators) can be seen in the zigzag line of the observed discharge, even when there are no rainfall events. The zigzag pattern is often caused by pumping stations operating during the day, lowering the water level, then stopping, letting the water level rise again during the night. The modelled discharge simulates a smoother line. This phenomenon can, for example, be seen in mid of April and mid of May 2010 (top panel in Figure 5-2). The zigzag discharge time-series smoothens out with the three-day moving average graph. Human control may also be the main candidate to cause the deviation between the modelled and observed discharge in November 2010. In November, there are two peaks in modelled discharge following a rain event, but the observed data do not show the event. This may be attributed to unrecorded activation of additional pumps by the water management to pump out the excess water during and before those peaks. The overestimation of inflow in dry periods might be due to unrecorded inflow to the system (or at least the data were not known to the modeller).

Table 5-1. Model performance for 2010 simulation

Performance criteria	Daily comparison	3 day moving average
NSE	0.52	0.62
PBIAS	10.80%	10.85%
RSR	0.69	0.62
r	0.81	0.87

Table 5-1 presents several hydrological model performance indicators for the modelled discharge of 2010. Based on the performance criteria, it can be concluded that the model gave an adequate performance. With NSE of 0.62, the model is in the range of satisfactory (0.50<NSE<0.65) as also the RSR of 0.62 (0.60<RSR<0.70). PBIAS of 10.85% hints that it is a good model (Moriasi and Arnold 2007). These performance metrics indicate that the model is of sufficient quality to serve as base model for the further experiments in this thesis.

Although the model might have been further improved by trying to optimise the parameter values, it was decided to keep the parameter values close to the ones provided by experts.

5.3 SIMGRO model validation

The discharge validation was used as the main validation of the SIMGRO model of Rijnland and was conducted for the year 2011. Secondary validation, for other variables than discharge, was conducted for different periods between 2010 and 2013 depending on data availability. The EO-based data, such as actual evapotranspiration, has been used as validation in the period from October 2012 to October 2013. The year 2010 was used as the period of ground water level validation. The main objective of validating the model is to compare the discharge from the four main pumping stations with the model results. The discharge validation was chosen as the main validation, because the discharge from, and inlet to, the Rijnland water system is the most important variable for the Water Board to control its water system. The primary objective of this control is to maintain target water levels in the storage basin.

5.3.1Main Validation

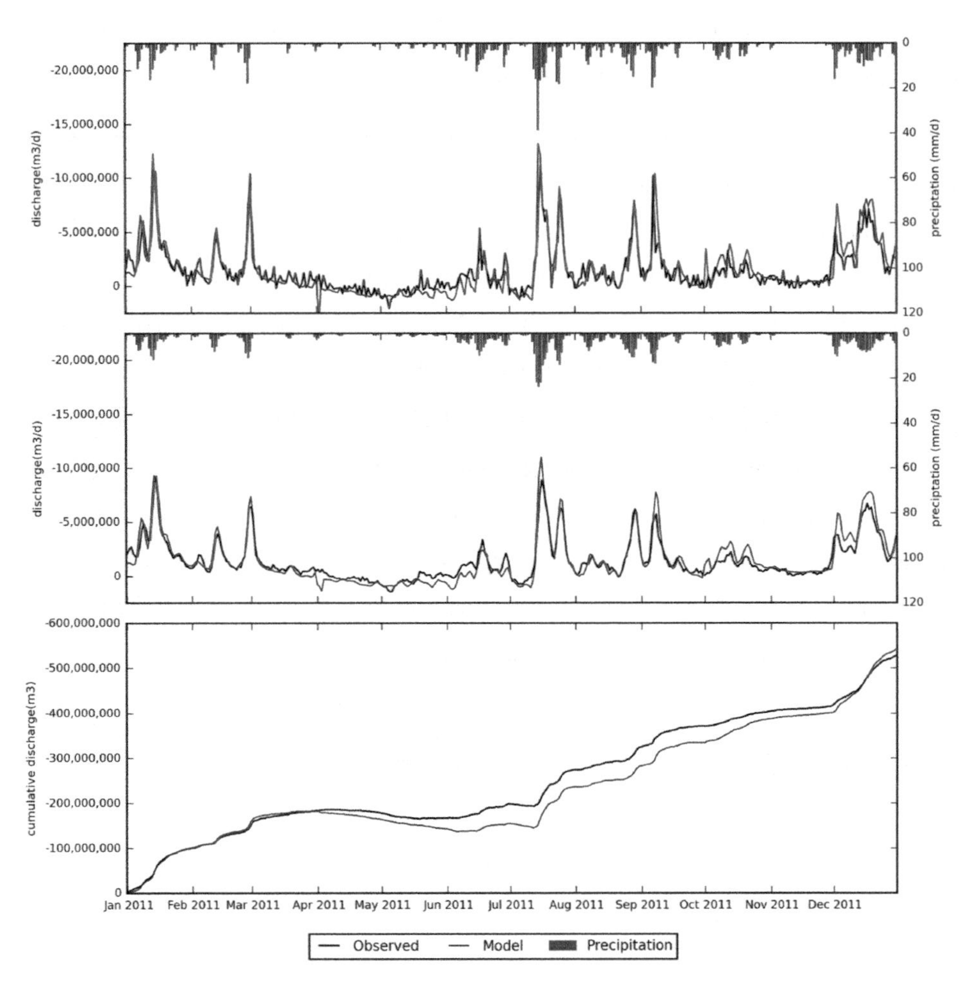

Figure 5-3. Discharge comparison of the base model, , the top graph is the original daily time series, three-day moving average displayed in the middle graph and the cumulative in the third graph

The base model gave a good result for the 2011 discharge simulation. As seen in Figure 5-3 top panel, the conventional daily values show zigzag biases while no rain occurred, due to the control of the water system as previously discussed. The modelled daily discharge does not include this zigzag pattern. The three-day moving average discharge shows very good performance of the model. Remember that the negative sign of the discharge values means that the water flows out of the system.

The results showed that the modelled discharge underestimates the measured discharge from April to June where the inflow in Rijnland was high. On the other hand, for the period from September to December, the model results showed the modelled discharge overestimate the measured discharge values.

The performance indicators showed that the model performance was rated satisfactory, even for the conventional daily comparison, as seen in the Table 5-2. The performances of the conventional comparison were in the "upper good" value range for NSE and RSR criteria, according to Moriasi et al. (2007). Moreover, the PBIAS is in the "very good" model performance with a value less than ±10%. Furthermore, the three-day moving average, which is used to minimize the human influence in the controlled water system, gave better performance ratings, most probably, since it smoothened out the noises. Noises occurred daily when the pumping stations pump more than what was needed and less in the day after, and vice versa. All performance ratings for the model in three-day moving average were in the very good rating. The PBIAS does not improve as much as other performance indicators. This inferred that the PBIAS performance criterion does not measure errors if they are cancelled out at another time step.

Table 5-2. Performance of the base model for the validation period (2011)

Performance criteria	Daily comparison	3 day moving average
NSE	0.73	0.85
PBIAS	-2.59	-2.44
RSR	0.52	0.38
r	0.89	0.95

Comparing the simulation results and observed data in the year of 2011, it shows that the base model performed well in simulating discharge. The performance for both original daily time step and three-day moving average was satisfactory. Furthermore, the model result of three-day moving average time series, where operator influence is largely filtered out, indicated a very good performance score. Note that the performance of the model during validation turned out to be higher than the performance for the model preparation period (2010).

5.3.2 Secondary validation

The available observed data used for secondary validation were local polder discharge, actual evapotranspiration, soil moisture, ground water level, and surface water level.

Not all of these additional measured data sets are of highest quality. The EO Actual Evapotranspiration and soil moisture data for instance, have the highest uncertainty amongst these data. The uncertainty comes from the nature of EO data where the measurement of the variables is done indirectly (Schmugge et al. 2002) and might differ between regions and time periods. Similarly, the discharge data of local polders was derived from the pump operation data (on-off), which also introduces uncertainty into the data, although not as much compared to EO data. The surface water level was measured with high precision, however, the model was not built to simulate the water level accurately. Thus, these secondary validations could be called a "mutual validation", where both observed data and modelled data cannot simply be accepted as the ground truth. Both data might be inaccurate. The main validation was done for data of 2011 using the discharge from the four main pumping stations, while the secondary validation was done using multiple data sets for the periods when the data was available.

Validation against the observed pump discharge in a local polder

Validation against the discharge of a polder was conducted for two polders, Vrouw Vennepolder and Zilkerpolder (Figure 5-4), for the year 2011 and 2012. The validation was conducted for cumulative discharge instead of daily time series, due to the uncertainty in the time series data. The Vrouw Vennepolder is located in the central part of Rijnland, near the city of Leiden. The polder has average elevation of -2.00 m+NAP with peat soil. The Zilkerpolder is located near the dunes area just south of De Zilk area with elevation around -0.30 m+NAP and has sandy soil.

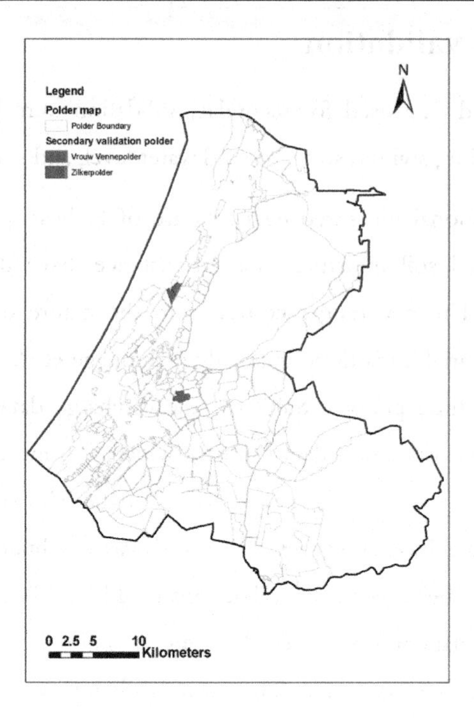

Figure 5-4. Location of polders for secondary validation, Vrouw Vennepolder and Zilkerpolder

The comparison between simulated and observed discharge from Vrouw Vennepolder is presented in Figure 5-5. For the first six months in the validation period, the model simulated discharge matches with the observed data. However, during the summer dry periods the model underestimated the discharge, especially around July and August 2011. The cumulative discharge graph coincided again around February 2012. In July 2011, the model simulated almost no discharge pumped out from the polder, while the observed data reports that there was outflow from the polder. The difference might be due to unrecorded inflow into the polder (inlets), causing the pump to pump some water out. The inflow could come from the flushing that has been carried out in the polder at that particular time. Another possibility is that farmers, expecting a very dry period, might have opened the inflow gate to maintain the water level in the area. In contrast, in 2012 the model simulated an overestimation of the discharge, even in summer periods.

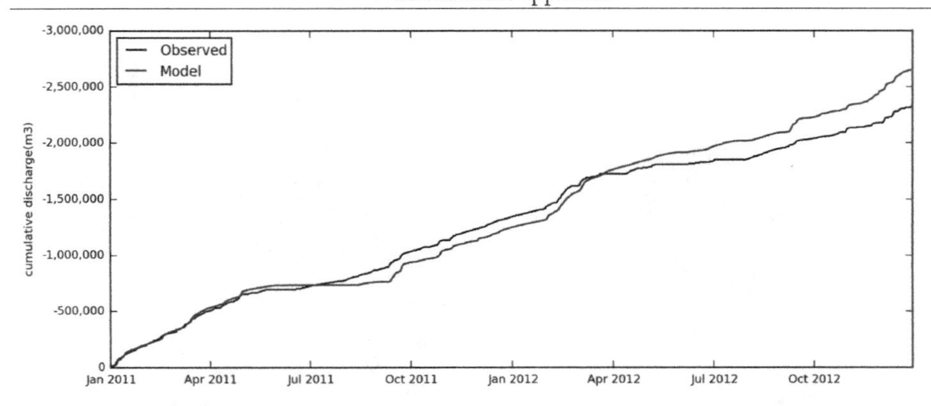

Figure 5-5. Validation of simulated discharge in Vrouw Vennepolder

For the validation in Zilkerpolder, the comparison can be seen in Figure 5-6. The model was able to simulate the discharge with a good comparison against the observed data. Although in 2011, the model simulated higher discharge for the beginning of summer period and lower discharge in the winter at the end of the year. However, looking at the year 2012, the model underestimated discharge in the beginning of winter period in September.

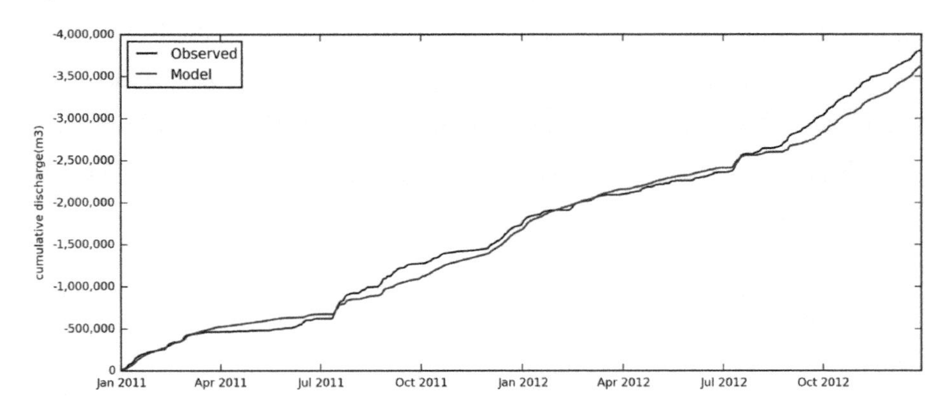

Figure 5-6. Validation of simulated discharge in Zilkerpolder

Figure 5-7 presents the total simulated flow of water into the grid cells from the canals, including the runoff, overflow, infiltration, and drainage. Obviously, the open water in the main storage basin, as seen as the red areas, contributes the most water to the grid cells. The low-lying areas contribute most water to the canals.

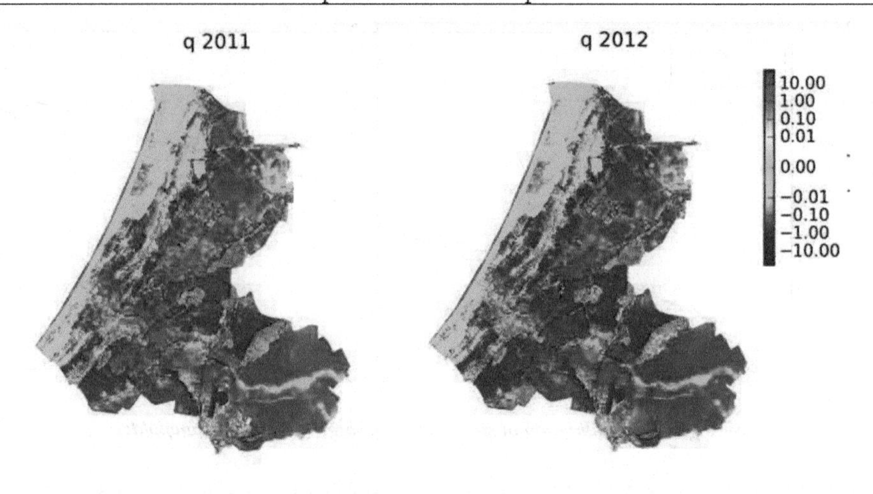

Figure 5-7. Modelled cumulative flow (in meter) from the canals to the grid cells (runoff and drainage) in 2011 and 2012. (Positive indicates water is flowing into the cells)

Validation against actual evapotranspiration from EO data

The validation of ETa was conducted by comparing the calculated ETa from SIMGRO model with the ETa from EO data. The source and characteristic of the EO ETa data were explained in Section 4.5. There are two types of EO ETa maps available, the composite daily ETa maps, and the eight-day cumulative maps. The eight-day cumulative evapotranspiration data was more reliable than the daily composite data. The daily composite maps had more missing data.

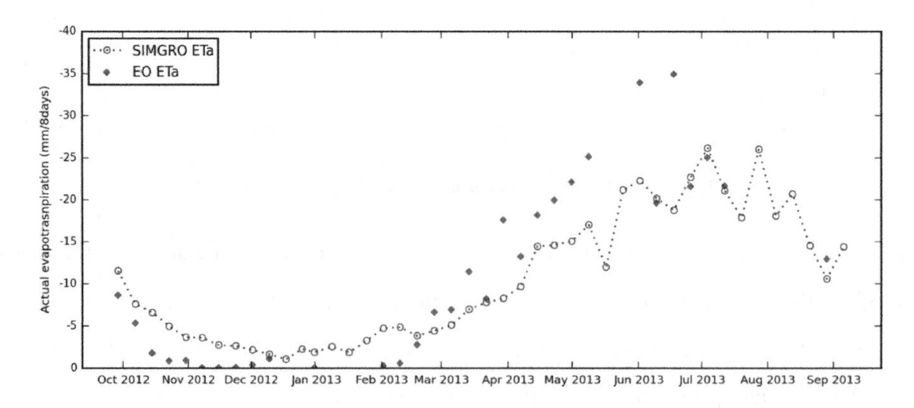

Figure 5-8. Comparison of eight day cumulative ETa, area averaged through Rijnland

There are three validations that were conducted for ETa: the averaged ETa of Rijnland area, ETa at a point location, and frequency analysis comparison. The comparison can be seen in Figure 5-8, where the eight-day cumulative area-averaged ETa from the model is compared to the averaged eight-day cumulative from EO data. Generally, the model gave a good comparison of simulated ETa to the EO data. However, the model simulated higher ETa at the beginning of winter and lower ETa in the end of winter and beginning of summer.

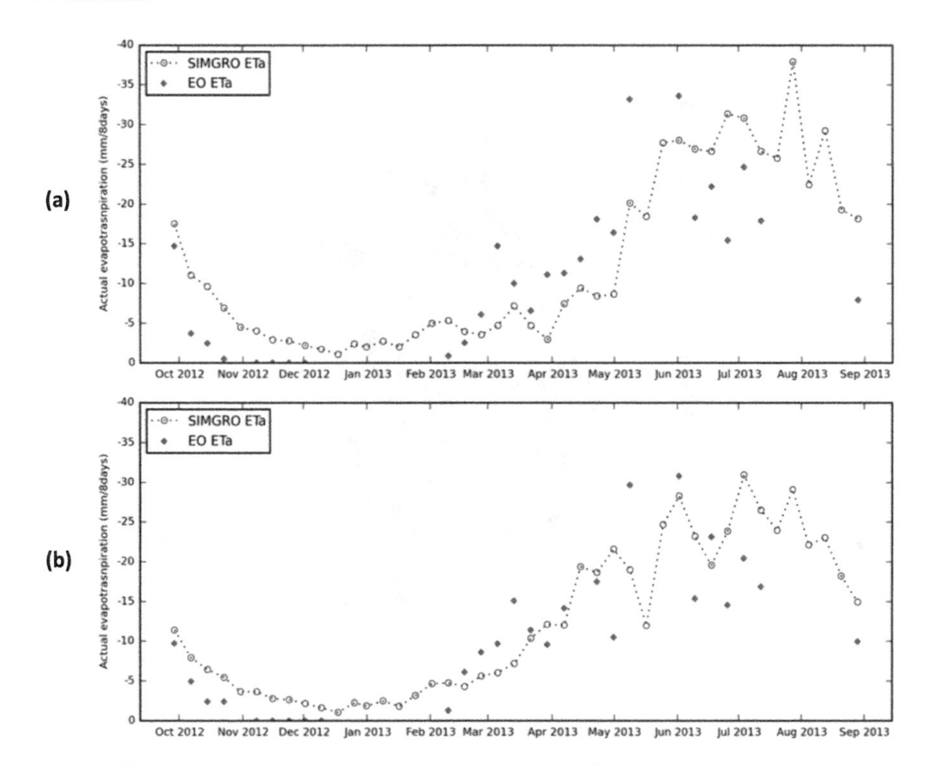

Figure 5-9. Eight-day cumulative ETa simulation from EO and SIMGRO model for one grid cell with deciduous (a) and grass (b) land-use

Looking at point locations, the comparisons illustrated similar pattern to the area-averaged comparison. Both of the point locations with deciduous and grass vegetation, presented in Figure 5-9, show that the model simulated higher ETa in winter and lower ETa in early summer. However, on the contrary to the area average, the model resulted in higher ETa compared to the EO in the late summer period.

The point locations for daily time step comparison were selected based on the land use in Rijnland. The twelve location points can be seen in Figure 5-10, with four land use types each in north, centre, and south part of Rijnland. The three main land uses for these points are herbaceous crops (211, 212, 213), trees or forest (311, 312, 313, 321, 322, 323), and non-crops herbaceous vegetation (351, 352, 353). The location was also selected as it has a consistent land use in 250 m in radius to fit the resolution of the EO data, which is 250 m as well.

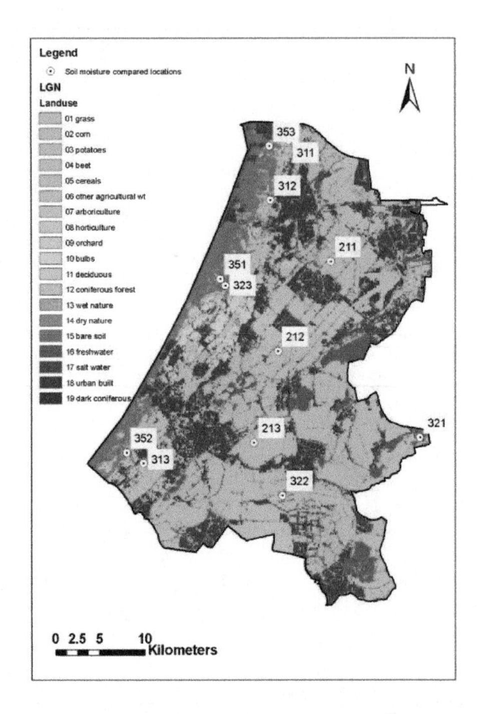

Figure 5-10. Locations of the ETa and soil moisture point comparison for daily time step

The daily ETa comparisons are presented in Figure 5-11, comparing the EO ETa composite map with the continuous time series of SIMGRO model results. The difference in resolution needs to be noted, the EO data used 250 m and SIMGRO model used 50m. Observing the figure, the modelled ETa has the same general pattern with the EO data. However, the model's over estimation in winter and under estimation in early summer remains noticeable.

Figure 5-12 shows the frequency distribution of ETa for summer periods, April 2013 to September 2013. From the frequency distribution view, the model calculated more low ETa values of 0 to 2 mm/day and simulated less middle and high ETa values. However, the pattern and distribution between the EO and modelled is similar. The difference in the ETa values might come from the errors in both data.

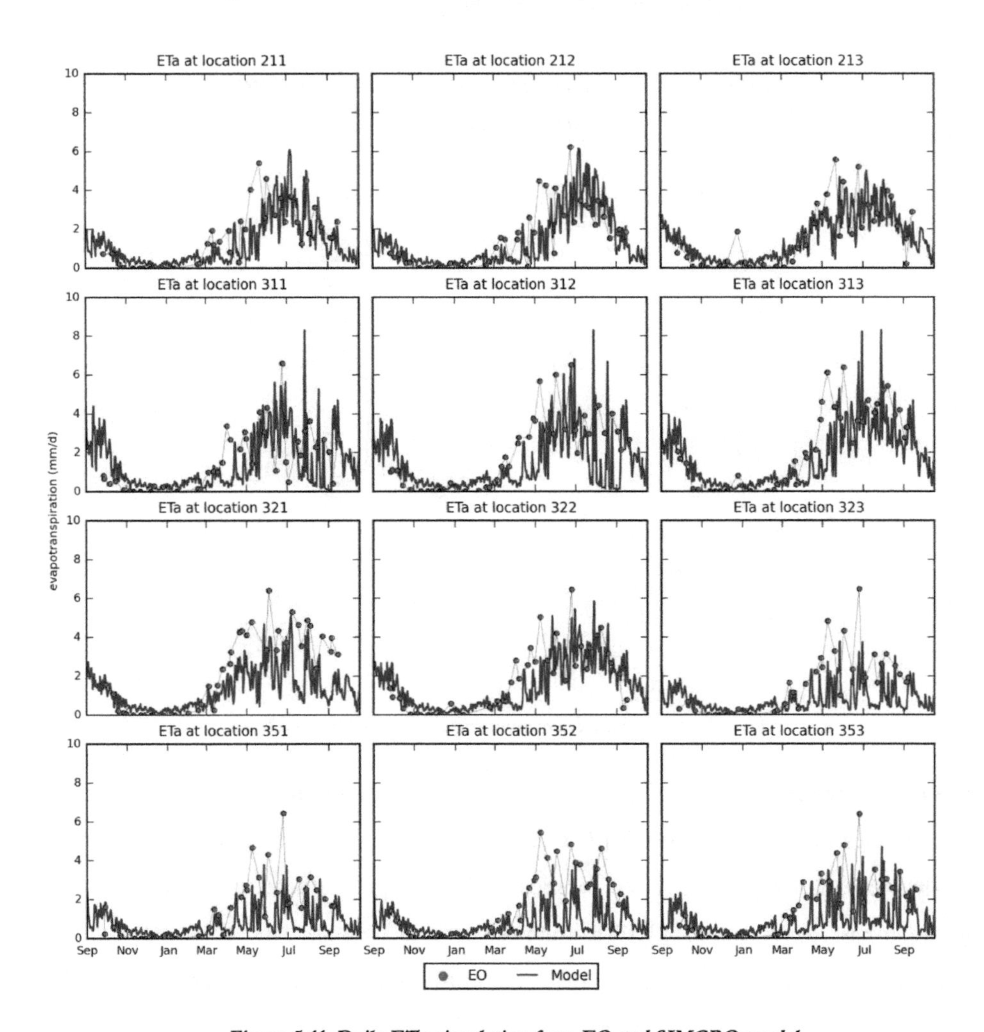

Figure 5-11. Daily ETa simulation from EO and SIMGRO model

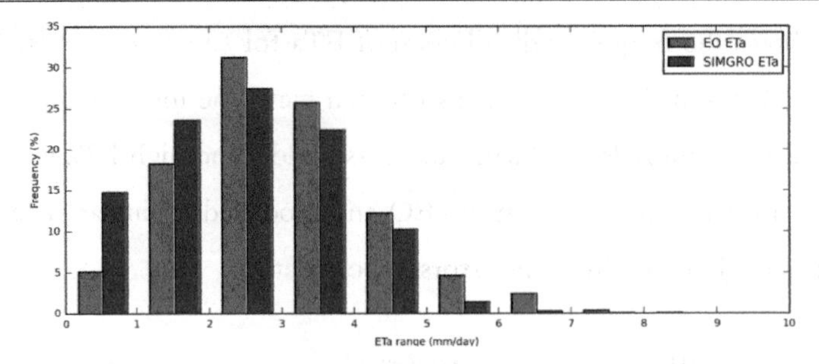

Figure 5-12. Frequency distributions of ETa values from EO and SIMGRO model (April to September 2013)

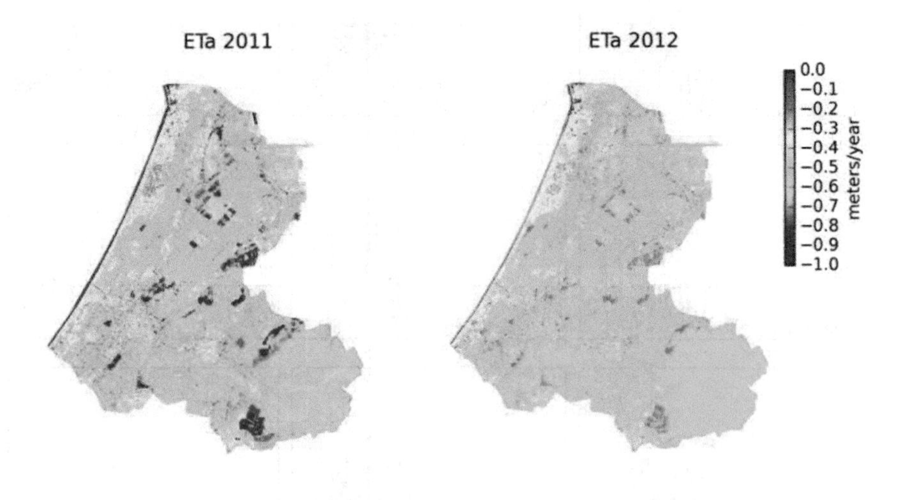

Figure 5-13. Cumulative evapotranspiration in 2011 and 2012

The cumulative ETa map in Figure 5-13 shows that the high evaporation occurs in the water bodies in both 2011 and 2012. The grassland gives second highest ETa, with the low-lying area having higher ETa compared to the high ground level area. The urban area generally has lowest simulated ETa, followed by the bare soil and forest. The model shows that the simulated ETa highly depends on the land-use type. The ground elevation was also affecting the ETa variability, which is linked to the ground water level.

The secondary validation with ETa shows that the SIMGRO simulation and EO data generally agree for Rijnland. Slight variation was noticed as EO ETa has a higher ETa in summer and lower ETa in winter compared to the SIMGRO model. The ETa pattern throughout the season is similar. The one grid-cell analysis also shows the same conclusion. The frequency analysis of summer period shows that the SIMGRO model simulated more cells with lower ETa than the EO data.

Validation against soil moisture from EO data

Another EO data, i.e. soil moisture data, was available to validate the model. The root-zone soil moisture data was derived from ETa data as presented in Section 4.6. The data was available in eight-day composite map, the same as ETa data. However, there is no cumulative data since soil moisture is a state variable. The soil moisture validation was conducted in several methods similarly to the ETa validation: soil moisture at a point location, and frequency analysis comparison.

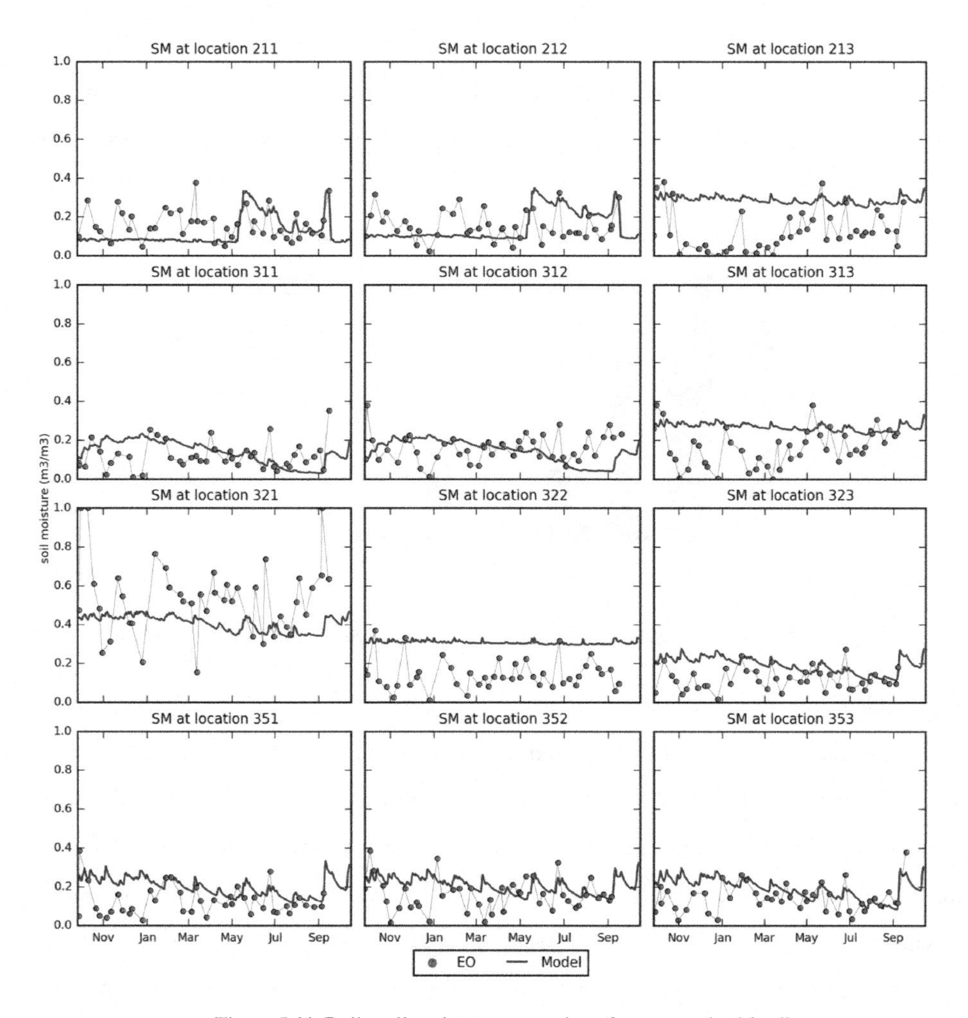

Figure 5-14. Daily soil moisture comparison from several grid cells

The point locations comparison of EO soil moisture and modelled soil moisture can be seen in Figure 5-14, where several locations with different land use were compared. The graphs show that both data sets were mostly not in agreement, although for some periods a similar pattern occurred. For instance in the peak event in September 2013, both data gave almost the same soil moisture at location 211 and 212. However, such similarity was not found for location 311 and 312. Furthermore, in the winter periods, in October 2012 to April 2012, there were almost no similarities between the two data sets, neither in pattern nor in soil moisture values.

Looking at the frequency distribution of soil moisture in Rijnland, the EO data and SIMGRO model further deviate from each other, as presented in Figure 5-15. The EO data tends to simulate low soil moisture values, with more than 80% of grid cells having less than 0.20 m^3/m^3 of soil moisture from April to September 2013. While the SIMGRO model simulated more than 70% of the grid cells having soil moisture higher than 0.20 m^3/m^3.

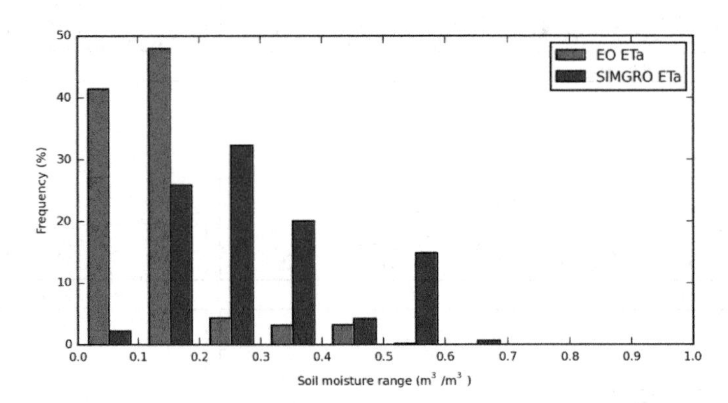

Figure 5-15. Frequency distributions of soil moisture from EO and the SIMGRO model

To be able to determine which data set has more correctness in this mutual validation, an additional analysis with other data is needed. Precipitation is considered directly affecting the soil moisture content, and the precipitation data has high confidence level. Hence, the comparison of precipitation pattern with the soil moisture pattern was utilized for the analysis. Figure 5-16, Figure 5-17, and Figure 5-18 presents the comparison between precipitation data and both soil moisture data sets. The comparison shows several moments of incorrect behaviour of the soil moisture from EO data, where a precipitation event occurred but the soil moisture was getting lower. These deviations might be due to the low quality of EO data during high precipitation with high cloud coverage. Several locations show reasonable pattern comparison.

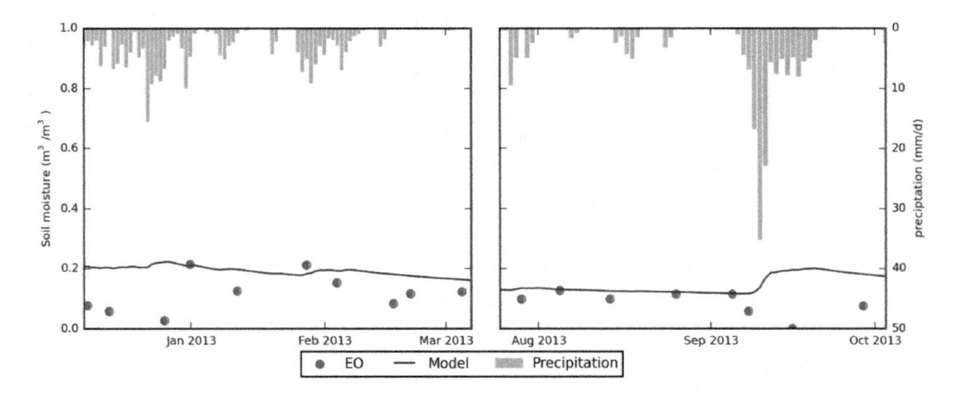

Figure 5-16. The comparison of precipitation events and soil moisture pattern of a cell with sandy soil in northern Rijnland.

The soil moisture validation did not give satisfactory results, as the two data sets (EO and model) show strong deviations from each other. In absence of direct measurements of the soil moisture in the Rijnland area, it is not known which data set has a higher correctness. However, when comparing the soil moisture from EO data with precipitation, these have different patterns. Hence, the soil moisture data from EO were not considered good enough and were not used in further analyses of this research.

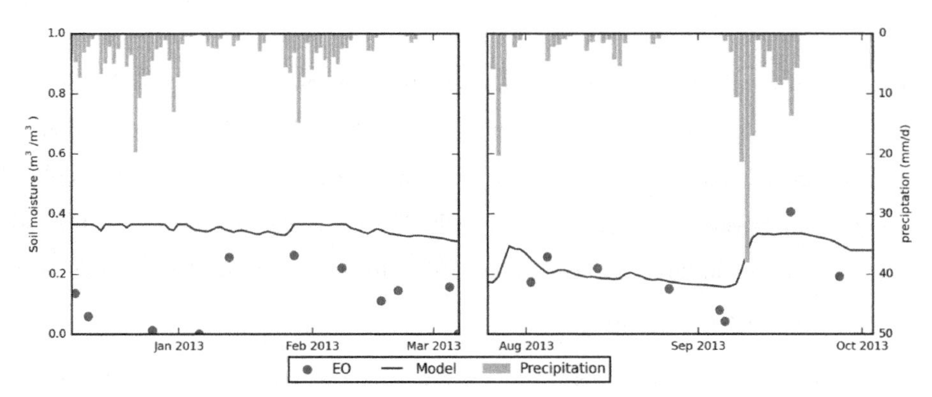

Figure 5-17. The comparison of precipitation events and soil moisture pattern of a cell with clay soil in southern Rijnland.

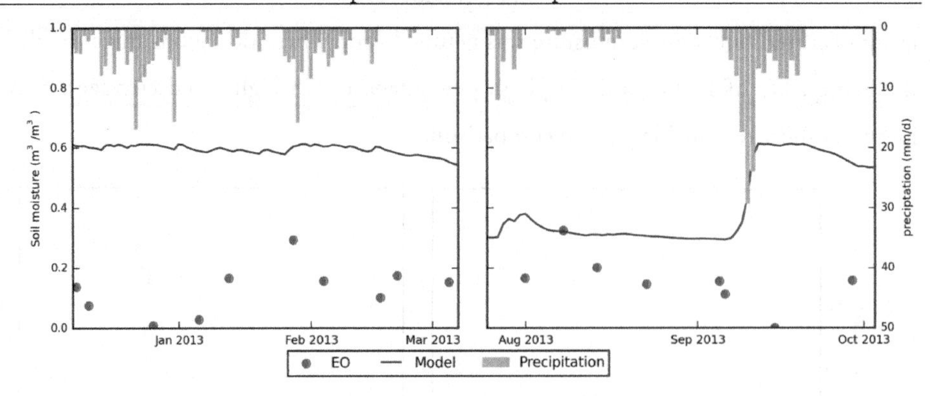

Figure 5-18. The comparison of precipitation events and soil moisture pattern of a cell with peat soil in central of Rijnland.

Validation against surface and ground water level

The observation of surface water level in Rijnland's canals was quite accurate with 10 minutes temporal resolution data from canals in the polders. However, the SIMGRO model is not build to model the sub-daily water level variations in the canals, such that a close match with observations is not expected. Large deviations are expected twice a year, because of the lack of data of occurred changes from summer to winter target level and back. Different dates are chosen each year for when to change the target level from summer time to wintertime management. Furthermore, each polder might have its own time for level switching that differs for each polder and also differs from the main storage basin.

The ground water level observation data was also available for the Rijnland area as there are many ground water level measurement wells throughout Rijnland. The data came from Dinoloket (Dino loket 2013), which can be requested and available for public. The data availability varies depending on the wells. The latest data that could be obtained for this research was in 2010. Hence, the validation for ground water level as well as water level was conducted for 2010.

The location for the water level validation was selected based on available measurement wells in Rijnland and data availability in 2010. The selected point is located at Vierambacht polder in the central part of Rijnland. The ground water level data was obtained from a point in a field, while the surface water level information was obtained from the nearest canal. The comparison of SIMGRO results and observed data is presented in Figure 5-19. The graph shows a satisfactory ground water level comparison. The pattern of the ground water movement from observations and model simulation is comparable, although at some periods it deviated. The deviation could be seen in August 2010, where the observed data give a peak ahead from the SIMGRO model results. While in October, the observed data was lagging behind the modelled data. The surface water level comparison also shows a comparable result. Although a constant difference of 5 mm can be seen throughout the periods. The difference might be caused by either the inaccuracy of the target water level data being used in the model or a slip in the setting of actual target water level for the pumps in the polder. A time difference in switching target level was also observable, the observed data had earlier switch of target level

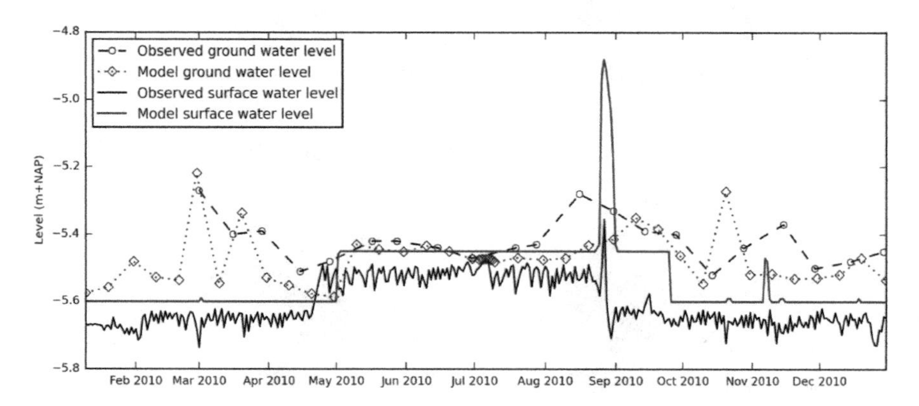

Figure 5-19. Ground water level and surface water level comparison at Vierambacht polder

In this research, the GXG map set from the NHI model result is utilized as one of the mutual validation for the SIMGRO model. The NHI model is an already accepted as a good hydrological model for the whole Netherlands. However, the NHI model has a five times lower resolution of 250 m compared to the 50 m of the SIMGRO model. The comparison of GXG map set from SIMGRO model and NHI model is presented in Figure 5-20. The comparison generally shows agreement in ground water level pattern across the modelled area. The significant difference can be seen in the water body, especially in the main storage basin. The difference is due to the different DEM data of the bottom of the water body such as lakes and wide canals. The SIMGRO model has a more detailed representation of the ground water level due to the higher grid resolution, as seen in the Schiphol area, a low area in North West. The resolution difference was not only affecting the level of detail of the ground water level map, but also the target water level and canals modelled in the grid cells, which strongly influence the ground water level calculation. The GLG maps have the least differences with an average bias of 0.03 m difference while GHG has 0.07 m average difference and the GVG has the largest average bias of 0.20 m. Explanation about GXG, GLG, GHG and GVG maps can be seen in Section 4.6.

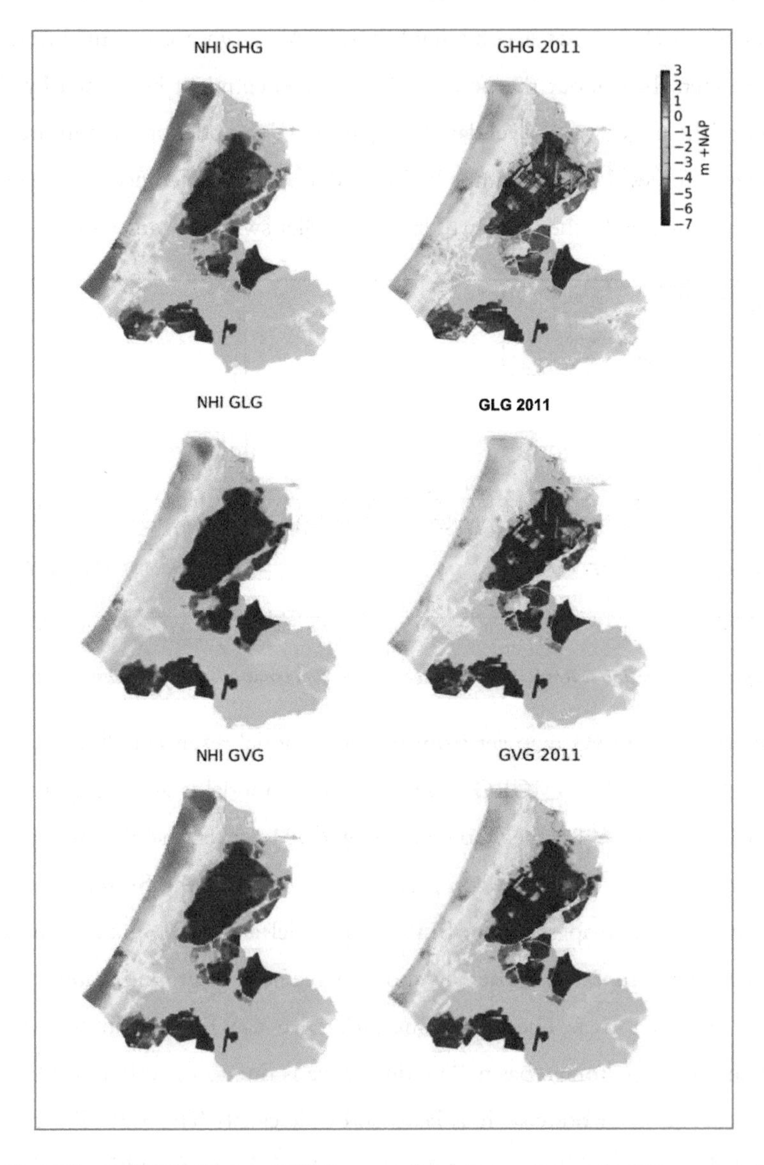

Figure 5-20. GXG map (GHG, GLG, and GVG) comparison of ground water for 2011 compared to the map of NHI

5.4 Uncertainty analysis

Parameter uncertainty analysis was also performed. Infiltration and drainage resistance of the canals, micro-storage capacity of the ground surface, and infiltration rate of the soil surface were chosen as the parameters in this analysis, because of high uncertainty when estimating the values for these parameters or because of expected high sensitivity of the model results to these parameter values. This selection was based on expert judgement from the Water Board. The sampling was not done spatially, but instead it was obtained by multiplying the variables with a factor for the whole area.

Table 5-3. The sampling method, parameters for the normal distribution

Variable	Mean	Sigma
In out resistance	25 days	12.5 days
Micro storage capacity	Depends on land use (0.01 to 0.02 m)	0.5 x
Infiltration rate	0.05 m/day	0.025 m/day

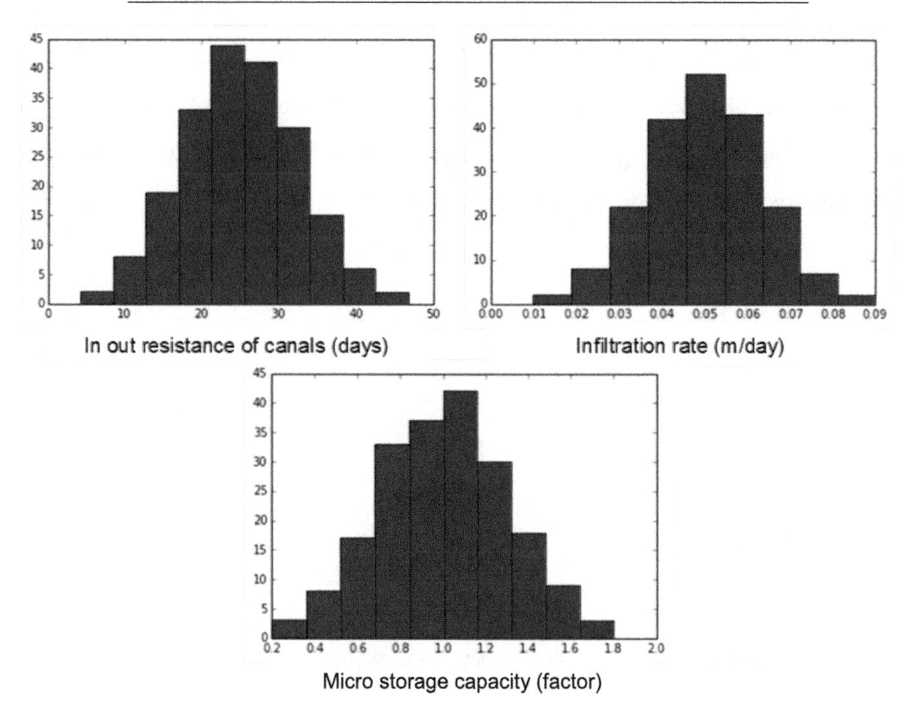

Figure 5-21. Distribution of sampled parameters, 200 samples

Sampling of variables was done with Latin Hypercube Sampling to ensure the tail in the sampling distribution has enough samples, hence lowering the number of samples needed. To keep sample count relatively low is important in this analysis due to the fact that the model requires high computational time. The variables were sampled with normal distribution. The mean of the distribution was coinciding with the parameter's value of the base model. The standard deviation (sigma) of the distribution was chosen to be 50% of the values. The parameter for variable sampling can be seen in Table 5-3. There were two sampling experiments for uncertainty analysis done in this research: 200 samples were run for one year (2012), and 2,000 were run for one month period of July. The lower number of samples for the one year simulation was due to the high computing demand of the model. The calculations were done in the parallel computer system SurfSARA to reduce the calculation time (Section 3.6). The parameter sampling for the 200 samples can be seen in Figure 5-21.

The model results with 200 simulations gave discharge outputs within the range of a good performance. As seen in Figure 5-22, the models still have good NSE, PBIAS, and correlation to the observed data. The NSE ranged from 0.77 to 0.85. A model within this range is considered as a very good model. While the PBIAS of 14% to 20% is considered as a satisfactory to good model (Moriasi and Arnold 2007).

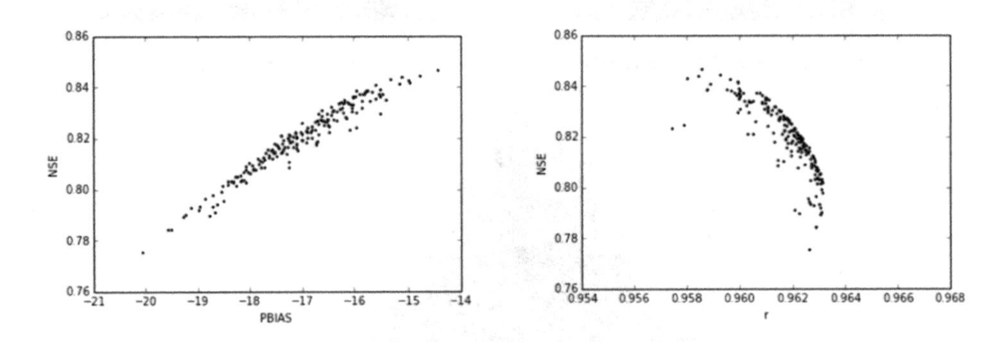

Figure 5-22. Performance of the model, for the 200 generated samples for 2012

The model performance with 2000 samples of the month of July, can be seen in Figure 5-23. For the July simulation, the model outputs with the 2000 samples were within the very good range for NSE scores, and the PBIAS scores are within the satisfactory to very good range.

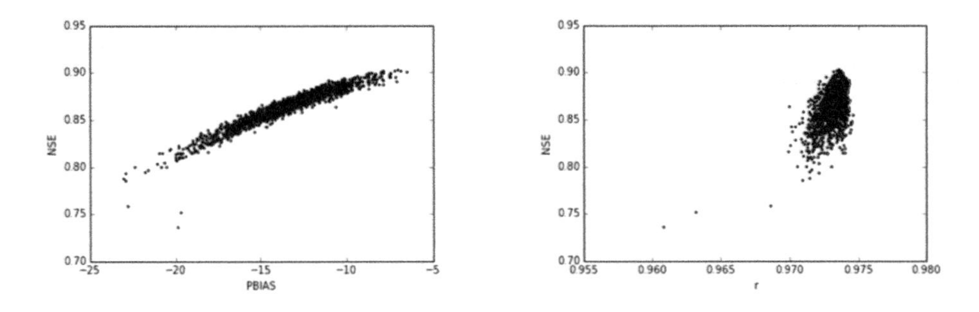

Figure 5-23. The performance uncertainty of the model, with 2000 samples for July 2012

The uncertainty analysis indicated that within the sample range of the parameters, the performance of the model was found to meet a good model criterion.

5.5 Summary

In building a good hydrological model, thorough understanding of the catchment is important. It becomes even more important in the case of a controlled water system, with additions of human influences and unnatural behaviour of the system. Good understanding is necessary to avoid accepting (seemingly) correct results for the wrong reason. Modelling of Rijnland showed advice from the experts in the catchment to be highly valuable. The expert advice included the operational rules and practice of the system, model parameter estimates, and which parts of the data could not be expected to be precisely matched by the model.

The SIMGRO model built in this chapter was able to simulate the total discharge of Rijnland with a satisfactory result. Although for the model set-up period the model performance was in the lower range of a satisfactory model, the model performed better in the validation periods. The errors were mostly observed during the summer periods when unknown inflow might have occurred.

Chapter 6. Data-model integration[1]

This chapter presents results of data-model integration methods applied to the Rijnland case study. Data-model integration methods are part of the overall framework to integrate multiple available data sources in hydrological modelling (central box in Figure 3-1).

6.1 Introduction

In this chapter, several data-model integration methods are applied. The first method is the direct use of available data sources for parameterisation of the hydrological model (when concerning catchment characteristics data) or as input to run the model (when concerning hydrometeorological forcing data). The second method is the merging of two data sources into one, in this case for precipitation, after which the effects on the hydrological model results are analysed. The third method is to use the hydrological model results for ETa to fill the missing data in the EO ETa maps and feed it back to update the hydrological model simulation (feedback loop). Lastly, data assimilation is applied to integrate the EO ETa data into the hydrological model.

6.2 Direct use of data in model parameterisation and simulation

Each of the data sources should be tested to, when used to build, parameterise, or force the hydrological model, result in an acceptable performance. Therefore, in this section,

---

[1] Parts of this chapter is based on Hartanto, I. M., van der Kwast, J., Alexandridis, T. K., Almeida, W., Song, Y., van Andel, S. J., & Solomatine, D. P. (2017). Data assimilation of satellite-based actual evapotranspiration in a distributed hydrological model of a controlled water system. International Journal of Applied Earth Observation and Geoinformation, 57, 123–135.

each of the data sources is used in the hydrological model developed in Chapter 5 to analyse the quality of the data. In Chapter 5, the base model parameter values were set. In this chapter, the model uses the same parameter values, however, there are parameters that are influenced by the data sources mentioned in Chapter 4, such as parameters related to land-use and soil type. As a result, when the hydrological model changes data source, the related parameter values are automatically altered.

6.2.1 Ground station and radar rainfall

Two precipitation data sources were presented in section 4.4, the rain gauge network and radar rainfall. The rain gauge network consists of 24 stations, spread in Rijnland area and Woerden. The radar rainfall data comprise 1-kilometre resolution precipitation maps. These two precipitation data sources are used as an input to the SIMGRO model to simulate hydrological processes in Rijnland.

Simulation of 2012 shows that the model with radar rainfall has a lower discharge bias to observed data than the model using the rain gauge network as seen in Figure 6-1. The rain gauges estimate higher precipitation than the radar in February and March 2012, as shown by the positive bias in Figure 6-1a. Figure 6-1c shows that the positive bias is reflected in the simulated cumulative discharge, where the rain gauges resulted in higher discharge than the radar data. Overall, the model using radar data provides better estimates of discharge than the one using the rain gauges, as presented in Figure 6-1b and c.

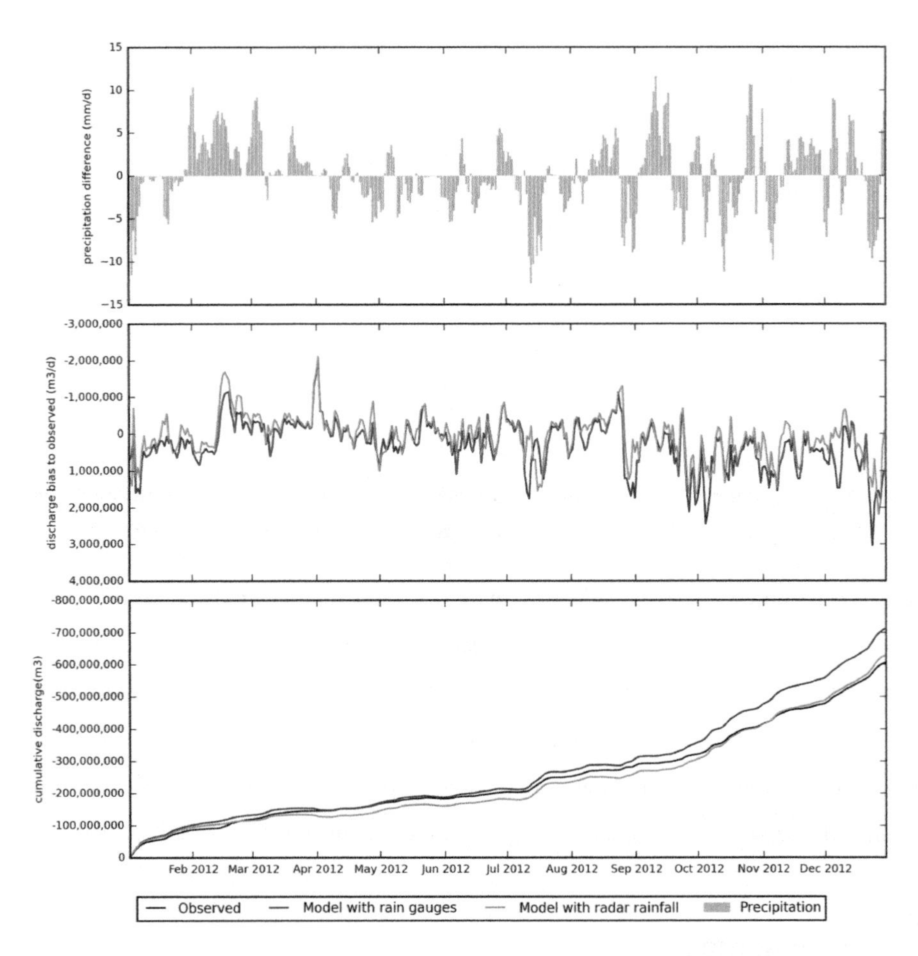

Figure 6-1. Main outlets discharge comparison of the model with rain gauge and radar to observed data for 2012.

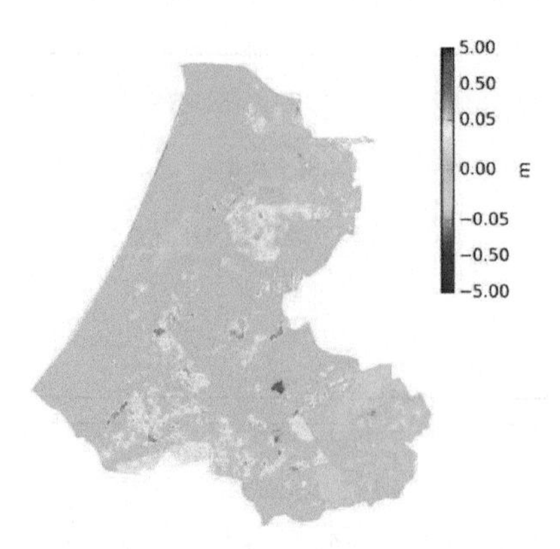

Figure 6-2. Spatial variability of the surface runoff difference in meters between model with radar and rain gauges for 2012. (Positive indicate the model with radar data simulates lower runoff)

Figure 6-2 presents the spatial variability in the differences in simulated runoff, with positive areas meaning the model with radar data generated higher runoff. The figure is for the year 2012 cumulatively. It shows that the dunes area in the west generates more or less the same runoff for both rainfall data sources. The model with radar data simulates lower runoff in southern Rijnland and in the centre part of the northern area. On the other hand, the radar rainfall generates higher runoff in the north and east, also in the Woerden area.

6.2.2 Land-use maps

As discussed in Chapter 4, there are three land-use maps which are used as the input to the hydrological model of Rijnland in this study LGN, LS5, and GBC. The land-use map is affecting simulation of several hydrological processes in the hydrological model, e.g. evapotranspiration, runoff, and infiltration. These processes in the end will be influencing the simulation of the total discharge from the Rijnland area.

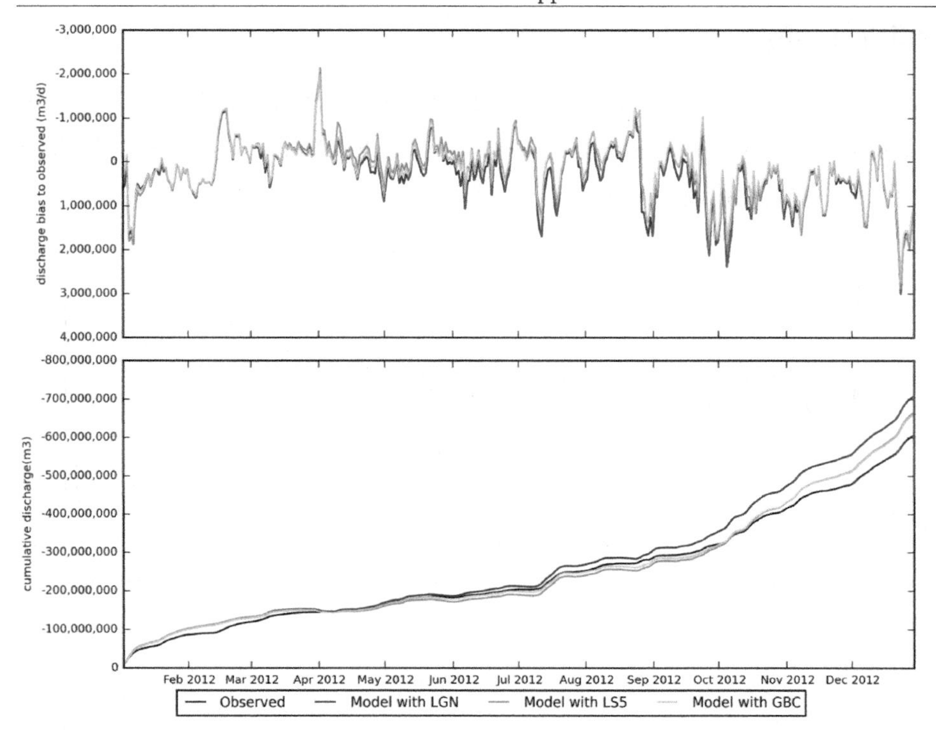

Figure 6-3. Discharge simulation results with the use of three land-use maps

Table 6-1. Performance of models with different land-use map (January to December 2012)

Performance indicator	LGN (Base model)	LS5	GBC
RMSE.STD.R	0.42	0.41	0.37
PBIAS	-17.06	-10.10	-9.39
NSE	0.82	0.83	0.86
r	0.96	0.96	0.96

The comparison of simulated total discharge in Rijnland system using the three different land-use maps are presented in Figure 6-3. The LGN land use is the one used for the base model. The influence of different land use inputs on simulated discharge is less significant than from different precipitation data discussed in previous section. However, the difference between the LGN and the two EO based data sets is observable, while the difference between the two EO based maps are insignificant in the cumulative discharge. Nevertheless, the two EO based maps resulted in an observable difference in the discharge in summer period.

6.2.3 Soil maps

There were two soil maps for Rijnland available for this study, the DDB and the EDB. These soil maps are explained in Chapter 4, and the EDB map was used in the base model discussed in Chapter 5.

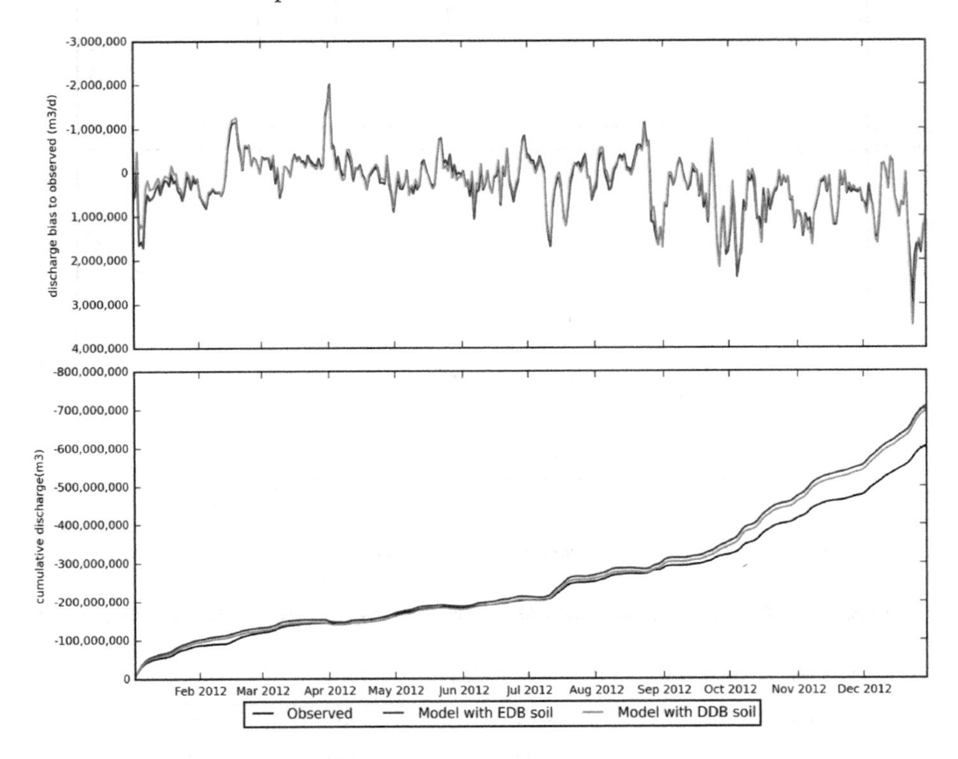

Figure 6-4. Discharge simulation comparison of the model with different soil map

When the two soil maps are input in the hydrological model, the simulated discharges do not differ much. The DDB soil map resulted in smaller cumulative bias to the observed discharge compared to the EDB soil map (Figure 6-4 and Table 6-2).

Table 6-2. Performance of the models with different soil maps (2012)

Performance indicator	EDB (Base model)	DDB
RMSE.STD.R	0.42	0.41
PBIAS	-17.06	-15.16
NSE	0.82	0.83
r	0.96	0.96

6.2.4Summary and discussion

The hydrological model results with different combinations of data sources for catchment parameters and hydrometeorological input, were presented in this section. The comparison analysis focused on the final discharge of four main pumping stations in Rijnland's main storage basin.

The two sources of rainfall, rain gauge and radar, resulted in different performance scores when analysed with the observed data. The radar data resulted in lower model performance in simulating discharge of 2011 compared to the rain gauges data. However, the radar data performed better in simulating 2012 discharge, showing that using both data sources can be beneficial.

The three land-use data sources led to smaller discharge model output differences as compared to the influence of the different precipitation data sources. The model with LS5 and the model with GBC simulated the cumulative discharge close to each other while the model result with LGN land-use deviated from the other two. Looking at the model performances for 2012, the GBC resulted in the best model performance although the map has the lowest resolution compared to the other two.

The two models with different soil maps, DDB and EDB, simulated very similar discharge for 2012. Nevertheless, the use of DDB leads to slightly better performance than EDB. The EDB model performs slightly better in the summer period.

The results show that, even if a particular data source, on average, results in better performance of the hydrological simulation, the other data source should not necessarily be discarded. The other data source may result in better performance in part of the analysis period.

6.3 Merging precipitation data from ground stations and weather radar

Precipitation is one of the primary inputs for hydrological model. With the availability of two precipitation data, there is an opportunity to merge the two data into one, before using as input to the SIMGRO model. Many studies have been done in merging rain gauge data with radar rainfall (Krajewski 1987; Seo et al. 1999; Goudenhoofdt and Delobbe 2009; Martens et al. 2013). The common merging approach is to use the rain gauge data for the calibration or validation of the radar rainfall (Tapiador et al. 2012), where the rain gauge data is used as a tuning tools for the radar. This kind of merging assumes that the rain gauge is more accurate than the radar data. Such merging is done by e.g. Sokol (2003), Velasco-Forero et al. (2009) and Martens et al. (2013). A comparison of different correction techniques has been carried out by Goudenhoofdt and Delobbe (2009). However, the quality of the rain gauge data itself must be checked before it is used in the radar correction (Vasiloff et al. 2009). Haberlandt (2007) use the radar data as additional information when interpolating the precipitation between rain gauge points. Sun et al. (2000) compared simulated discharge from hydrological model using rain gauge, radar and the merged data, they conclude that the merged data resulted in the most accurate result.

In this thesis, the radar data is already calibrated to the rain gauge measurement in Rijnland, although not all available rain gauges are utilized. However, the calibrated radar data still has biases compared to the rain gauge data as shown in section 4.4. Therefore, both of the data sets are merged through simple averaging by taking the average of the Thiessen polygon from rain gauge and radar rainfall for each 1 km grid cell. The merged data is then used as input to the SIMGRO model to simulate the total discharge of Rijnland.

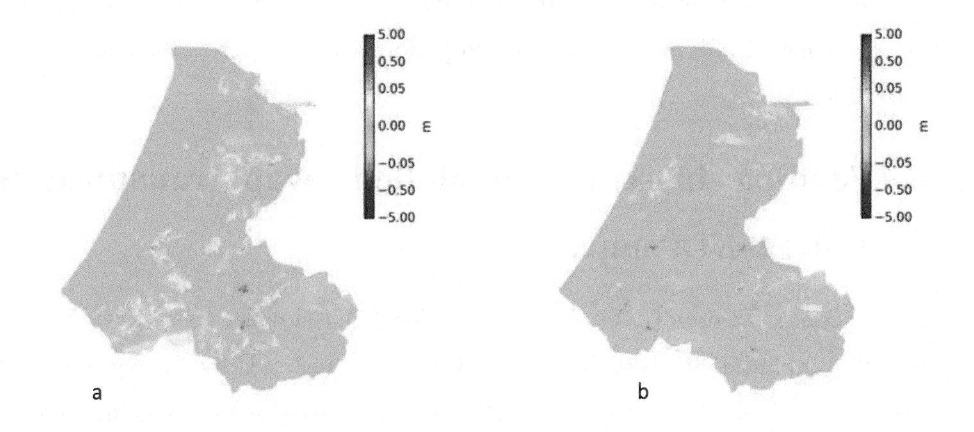

Figure 6-5. Spatial variability of surface runoff differences in 2012, the model with merged precipitation compared to the model simulation with Thiessen polygon (a) and the result with radar (b). (Positive means the merged precipitation resulted in a lower runoff)

Figure 6-5 presents the spatial variability in the difference of runoff generated by the merged rainfall compared to the rain gauge and radar data in 2012. The Thiessen polygon patches are visible in the runoff comparison between rain gauges and merged data, while in the comparison with radar these are not recognizable. The merged precipitation resulted in small differences in runoff for the dune area in the west, and locally in larger differences for the middle and eastern parts of the catchment. In general, the resulting runoff from merged data is in between the model with rain gauge and radar data.

Table 6-3. Performance of the model with different precipitation data source for 2011 and 2012

Performance indicator	2011			2012		
	Rain gauges (Base model)	**Radar rainfall**	**Merged rainfall**	**Rain gauges (Base model)**	**Radar rainfall**	**Merged rainfall**
RMSE.STD.R	0.36	0.41	0.38	0.43	0.36	0.38
PBIAS	-2.87	16.74	1.00	-17.59	-3.92	-10.76
NSE	0.87	0.83	0.86	0.82	0.87	0.85
r	0.96	0.95	0.96	0.96	0.96	0.96

Looking at the performance metrics in Table 6-3, the merged rainfall data has lower bias compared to both rain gauge and radar data for the 2011 simulation. However, the NSE and RMSR.STD.R score of the merged rainfall is in between the two model results of model with radar and rain gauge and slightly below the model with rain gauge input. In the 2012 simulation where the radar data resulted in higher performance over the rain gauge, the model with merged precipitation performance is closer to the one using the radar data.

6.4 Feeding back the modelled evapotranspiration into EO ETa maps

In this section, the use of ETa from EO (Section 4.5) as a direct input (forcing) to the spatially distributed SIMGRO model of Rijnland is analysed. By forcing a spatially distributed hydrological model to have ETa values equal to what EO data estimated, it is expected that the model performance would improve, at least in terms of the spatial distribution.

The hydrological model requires a complete set of data without no-data (empty) cells, hence, the no-data cells in the satellite's evapotranspiration maps need to be filled in. Furthermore, there are no available maps for some periods, for instance in early January due to heavy cloud coverage. Hence, the missing data are not only in space but also in time.

There are several interpolation methods to fill up the no-data cells or to improve the spatial data with point location data. Kriging, improved Kriging, and inverse distance weighting can be used for rainfall data (Haberlandt 2007; Nerini et al. 2015). The area-to-point Kriging was used in the downscaling of remote sensing data for ETa (Atkinson 2012). However, in this case study, the no-data problem is not only spatial but also temporal. Hence, hydrological model results are utilized to fill up the no-data cells in the ETa maps from EO.

The hydrological model is firstly run with reference evapotranspiration as input to calculate discharge and actual evapotranspiration. The no-data cells in the eight-day cumulative EO ETa maps are then filled in with the output of the first run of the hydrological model. The resulting EO ETa data are then used as input in the same hydrological model to calculate discharge again, thus creating a loop.

6.4.1Data updating and infilling using model result

The missing data of the EO ETa maps consist of the two types: the no-data cells in some maps, and missing maps for certain periods. Figure 6-6 shows the no-data cells in the available ETa maps, while Figure 6-7 presents the time where there are no ETa maps available.

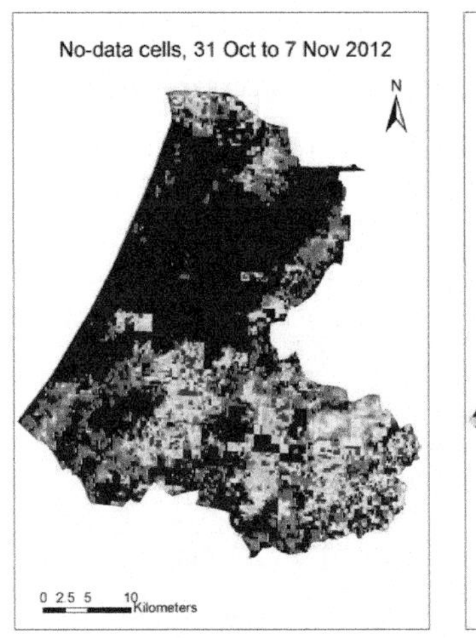

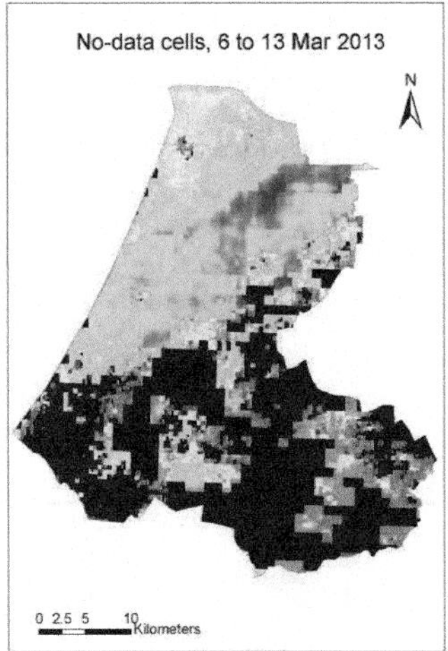

Figure 6-6. No-data cells (black) in ETa map of 31 Oct to 7 Nov 2012 and 6 to 13 March 2013

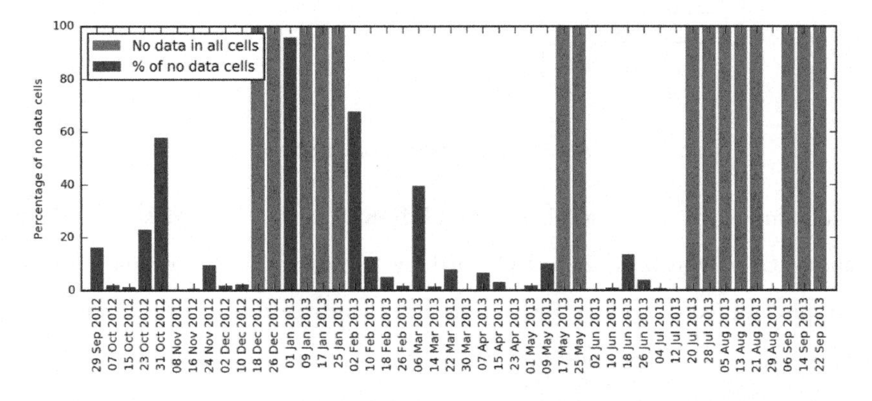

Figure 6-7. Missing maps and no data percentages of the EO ETa data set

There are 15 missing maps out of 46, each corresponding to eight-day cycles during the implementation period of 29th of September 2012 to 29th of September 2013. The average percentage of no-data cells amongst the available maps is 6 percent, 3.5% in summer period and 7.7% in winter period.

The flowchart of the data-model updating can be seen in Figure 6-8. The hydrological model was firstly run with reference evapotranspiration (*ETr*) input and other data as in the base model, calculated the actual evapotranspiration named ET_{out} using the data. The missing data in the EO *ETa* then filled using the calculated ET_{out}. The filled *ET* (ET_{filled}) was fed back into the hydrological model to calculate the final outputs. The process was creating a feedback loop to the hydrological model. The hydrological model and the EO data used different map projection and resolution, hence a spatial re-projection and resample was done. However, the resample is only necessary in the filling processes because the SIMGRO model is able to read a different map projection.

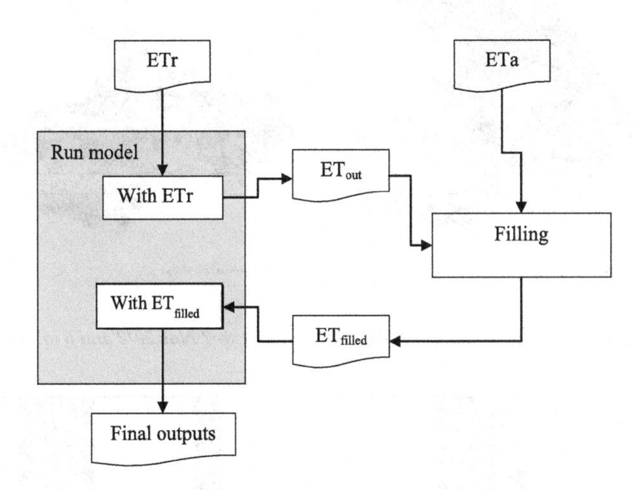

Figure 6-8. Flowchart of data-model updating

The default input for *ETa* calculation in SIMGRO model is a Makkink reference evaporation (*ETr*), which is then multiplied by a crop factor to produce potential evapotranspiration (equation 2.1). By altering all the land-use factors into a value of one, the calculated potential evapotranspiration will be equal to the *ETr* input. If the *ETa* map is used as the *ETr* input, the potential evapotranspiration will be equal to that from the *ETa* map. To have the calculated *ETa* the same as *ETr*, the moisture availability factor that limits evapotranspiration needs to be altered as well (equation 2.2). By implementing these steps, SIMGRO will calculate *ETa* which is equals to *ETr*.

With the above mentioned modifications, the filled EO *ETa* map can be used as the input to the SIMGRO model so that the simulated *ETa* is equal to the EO data. The simulated discharge is then compared to the observed one.

6.4.2 Data-model updating results

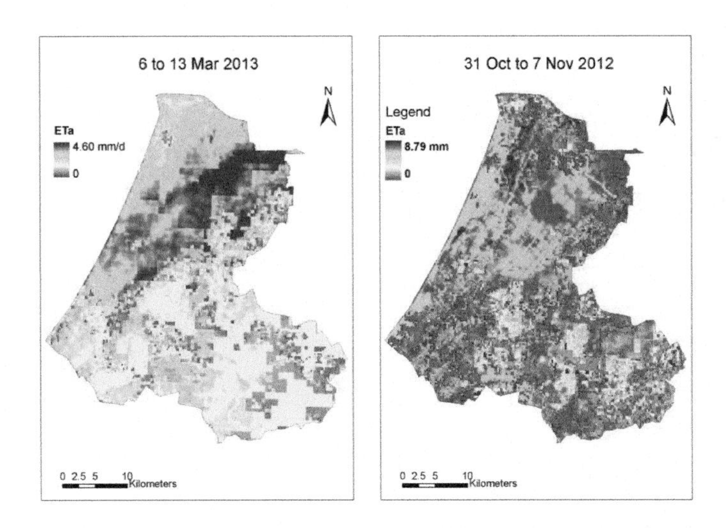

Figure 6-9. The filled EO ETa maps using the hydrological model simulation of Eta

Figure 6-9 presents the EO ETa maps that have been filled with the results from SIMGRO model calculation. ETa simulation shows that ETa results from model are the same as EO ETa, which indicates that the method is working. Figure 6-10 shows example of two ETa simulations from different point locations with grass and deciduous vegetation cover. In both figures, the calculated ETa is very close to the EO ETa estimations. The difference is negligibly small. Deviations are caused by the limitation of available water for ETa.

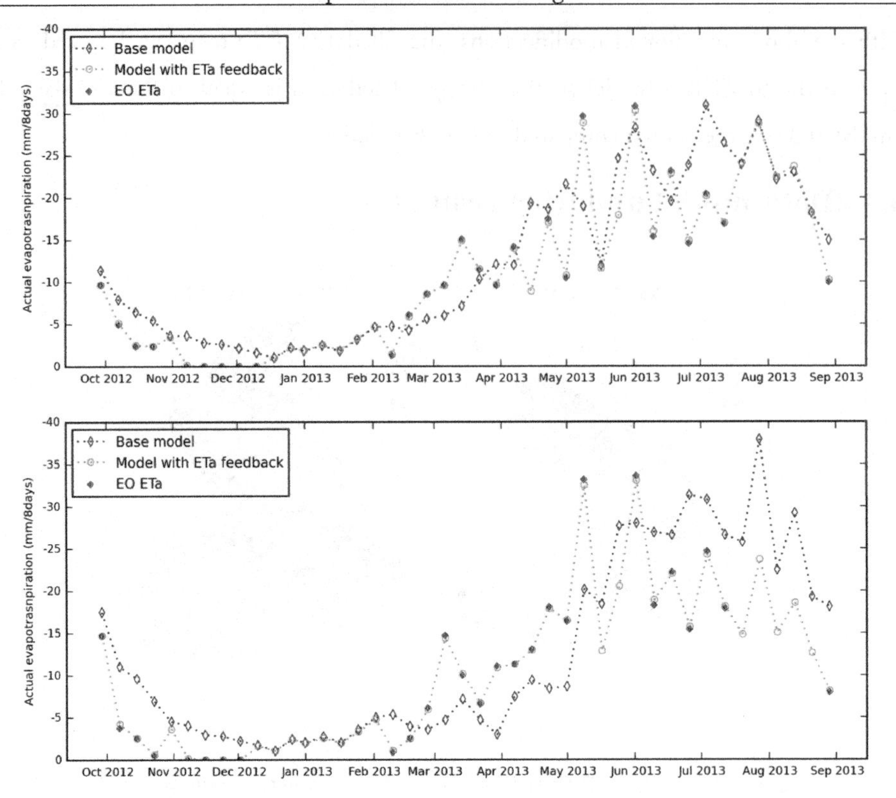

Figure 6-10. ETa comparison with the base model, on a cell with grass land-use (a) and a deciduous land use (b)

The discharge simulation from the model with ETa feedback loop can be seen in Figure 6-11. The result shows that the EO ETa feedback has reasonable results, with the peaks and low discharge reasonably simulated. The use of EO ETa data as input has little impact on the discharge simulation. This is shown by the small difference between the model results and ETr. Both models have a high difference to the observed data at some periods, e.g. overestimation in December 2012. As seen in the performance metrics in Table 6-4 for 2013, PBIAS of the base model is 21.0% in 2013, while for 2011 this was only 2.9%.

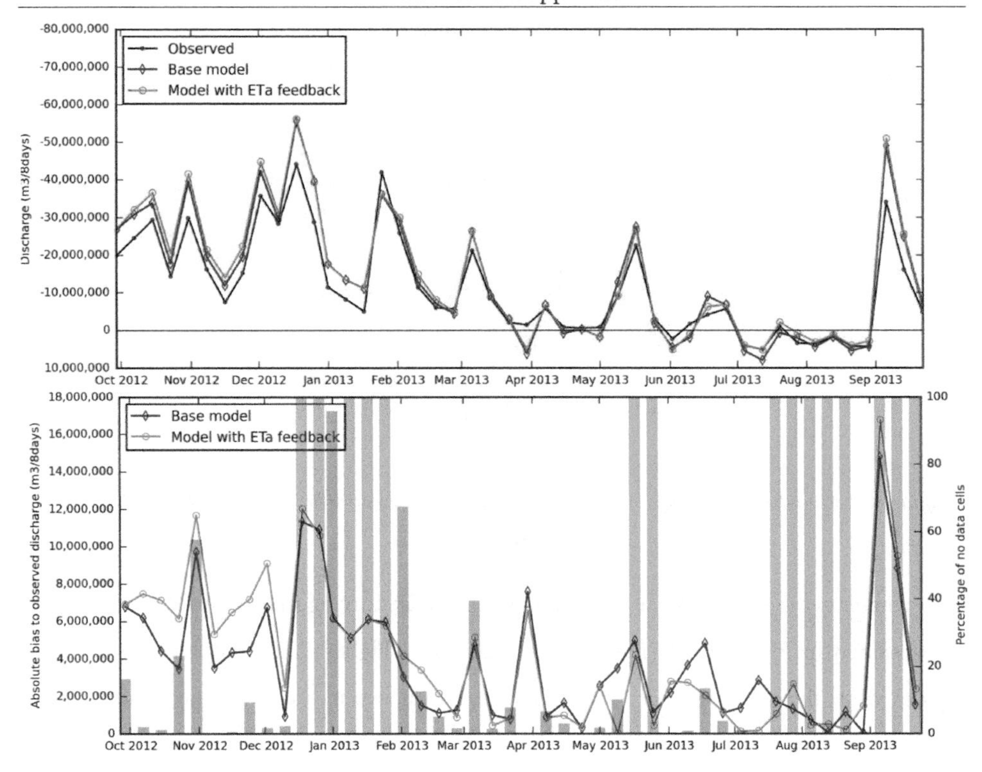

Figure 6-11. Discharge comparison of the base model with the filled ETa model

Table 6-4. Performance comparison of the modelled eight-day discharge, October 2012 - September 2013

Performance indicator	ETr (Base model)	ETa
RMSE.STD.R	0.39	0.43
PBIAS	-21.02	-27.91
NSE	0.85	0.82
r	0.98	0.98

The performance of model with ETa is lower than the one with ETr, but still within the very good model range for NSE and RSR (Moriasi and Arnold 2007).

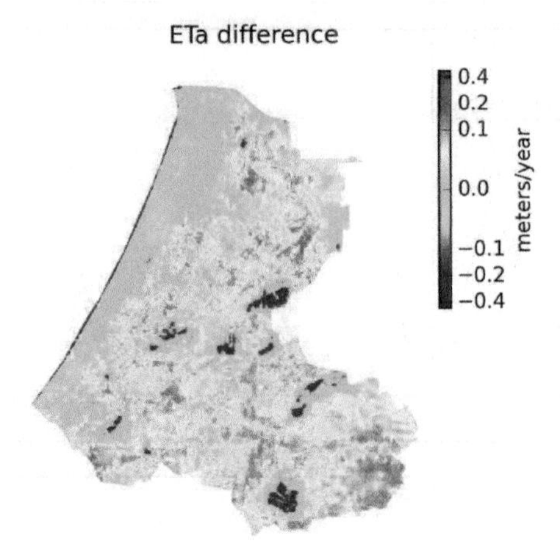

Figure 6-12. Cumulative ETa difference with the base model. (Positive indicates that the base model has less ETa)

The cumulative ETa difference map between the ETa forcing model and the base model is presented in Figure 6-12. The water bodies have higher ETa compared to the base model, while the grass land has lower ETa. The open area in the dunes has almost similar ETa.

The differences of the canals discharge generated by the grid cells are shown in Figure 6-13. The grass land and agriculture area generate less runoff and/or drainage to the canals compared to the base model. The water bodies have less infiltration to the unsaturated zone of the soil, and accept more runoff and/or drainage from the unsaturated zone. The runoff in the dune areas have similar runoff compared to the base model.

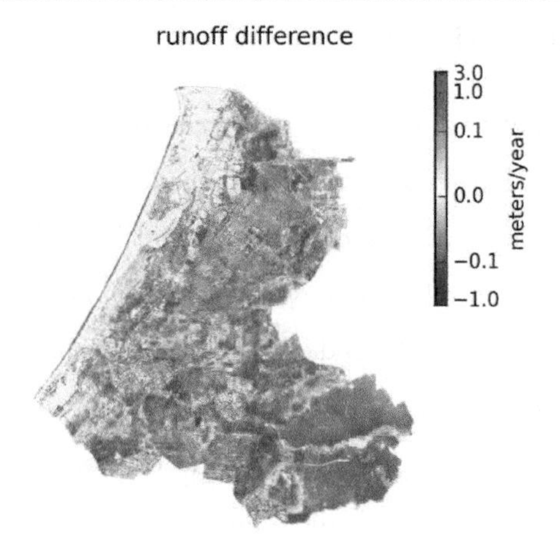

Figure 6-13. Difference in spatial variability of the simulated run off. (positives mean that the base model generate less runoff and/or less drainage to the canals)

6.4.3 Summary and discussion

The using of the model results to fill the no-data cells in the EO ETa maps and feed it back into the model presented in this section leads to a lower model performance in comparison to the base model. However, using the EO ETa maps improves the model simulation in several periods such as in parts of the summer period.

It is expected that by forcing the ETa from EO data into the model, the representation of spatial variability of the simulated ETa should improve, because the base model only uses one point location ETr and generates the spatial variability only through differences in land-use and soil moisture availability in the model. The EO data on the other hand, ensure the representation of ETa spatial variability, thus also including variation in, for example, meteorological conditions such as temperature and cloud cover. The model might need re-calibration to improve and utilize the EO data at the fullest.

Furthermore, the lower performance can also be explained by the long periods of missing ETa maps in the summer. It is in summer when, for the few periods that EO ETa was available, simulation results were improved.

6.5 Data assimilation

In the framework described in Chapter 3 and presented in Figure 3-1, data assimilation is one of the data-model integration methods applied in this thesis. Different from using the EO data as model input as presented in Section 6.4, the data assimilation implements a "softer" model alteration by accommodating uncertainties present in both model and EO data. The particle filter with residual resampling is chosen as the data assimilation method. The data assimilation is implemented by using open source software: PCRaster, and Python. The following sections describe the methods and software used in more detail, and then present and discusses the results for the Rijnland case study.

6.5.1Particle filter with residual sampling

Reichle (2008) and Liu et. al. (2012b) defined that data assimilation is to combine information from measurements and model results into optimal estimation of model states by interpolation and extrapolation in time and space. One of the popular methods for assimilation of remote sensing data in models is the Ensemble Kalman Filter (EnKF) (Evensen 1994, 2003), because it is flexible, accounts for nonlinearity, and has proven to be effective in many research projects on integrating data (Reichle et al. 2002; Clark et al. 2006; Moradkhani 2008; Xie and Zhang 2010; Crow et al. 2011). Another method is the particle filter.

Particle filters, or sequential Monte Carlo methods, are popular in the fields of object recognition (Zhou et al. 2004), target tracking (Hue et al. 2002), financial analyses (Yu 2005) and robotics (Fox et al. 1999). They have also been introduced in the field of hydrology (Moradkhani et al. 2005; Weerts and El Serafy 2006; Matgen et al. 2010; Dechant and Moradkhani 2011; Noh et al. 2011). The main advantage of Particle Filters is that no assumptions are made on the prior probability density function (pdf) of the model states and that the full prior density is used (Weerts and El Serafy 2006), while Kalman Filters assume a Gaussian distribution. With the sufficient number of samples, Particle Filters can approach the Bayesian optimal estimate and therefore have a wider applicability than the first-order approach of Kalman Filters.

The most common Particle Filters are based on either Sequential Importance Sampling (SIS) or Sequential Importance Resampling (SIR). SIS approximates the posterior density function of model states by a set of weighted Monte Carlo samples, called particles (Weerts and El Serafy 2006). The SIR algorithm (Gordon et al. 1993) is developed to avoid one particle having a very high weight while all other particle's weight being close to zero, which causes degeneracy of the algorithm (particle collapse). Resampling duplicates particles with a high weight according to their weight and ignores particles with a low weight (van Leeuwen 2003). Later on, further improvements of the SIR filter were developed. One of them being residual resampling (RR) filter (Liu and Chen 1998), chosen to be used in this research due to a smaller sample size, which leads to the lower computational time, just like the advantage of using SIR (van Leeuwen 2003). The Particle Filter with Residual Resampling method has been evaluated for discharge assimilation into a rainfall-runoff model and has shown its effectiveness (Weerts and El Serafy 2006). The Particle Filter schematization is shown in Figure 6-14. In case of a Particle Filter, DA is not an inverse problem which is good for maintaining posterior condition. The best set of particles is propagated, resulting in state predictions accompanied by its uncertainty, while preserving physical consistency, i.e. conservation of mass, momentum, and energy.

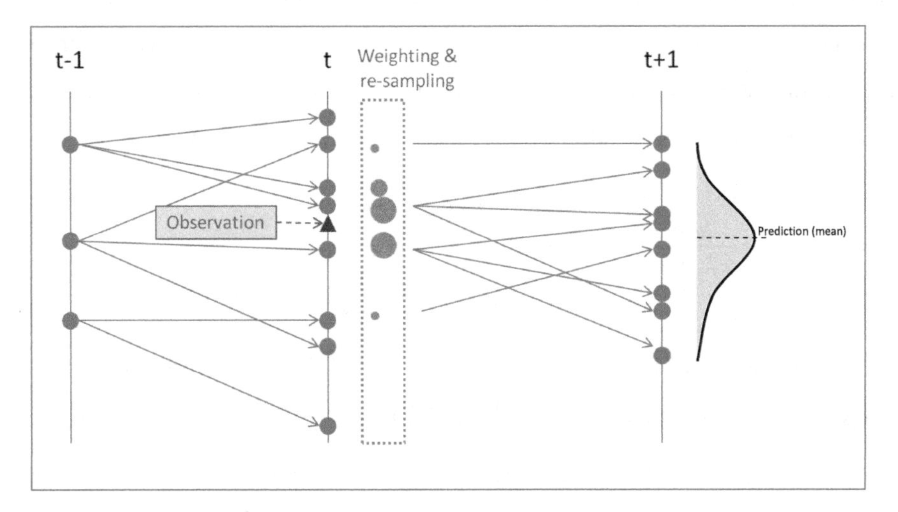

Figure 6-14. Particle Filter scheme. The blue dots are the particles. The ones with results close to the observed data are propagated to the next time step (green dots). The ones closest to the observed data are sampled to more particles (largest green dots).

In this research the resampling is done by multiplying the Rijnland SIMGRO model inputs with the same random factor for all the grid cells for each time step. Hence, the resampling of inputs is not done in a fully spatially distributed manner. However, the original inputs do already have spatial variability, which is retained after being multiplied by a factor.

The Particle Filter derives a posterior probability density function (pdf) by generating a set of Monte Carlo samples (model realisations, called particles) and using Bayesian statistics (van Leeuwen 2003; Karssenberg et al. 2010; Hiemstra et al. 2011). The Bayes theorem defined as:

$$P(y_i|d) = \frac{P(d|y_i)P(y_i)}{\sum_{j=1}^{n} P(d|y_i)P(y_i)} \tag{6.1}$$

where $P(y_i)$ is the prior probability of model realisation i, and $P(d|y_i)$ is the probability of the observation. $P(y_i|d)$ is the posterior probability or the weight of model i, or the likelihood of the model y_i is to the reality given by the observation. Assuming that the observation error has Gaussian distribution, the likelihood can be defined as:

$$P(d|\, y_i) = \exp\left(-\frac{(d-y_i)^T(d-y_i)}{2\,R}\right) \tag{6.2}$$

where the y_i is the model realisation vector, d is the observation vector and R is the covariance matrix of the observation error. Assuming that the observation errors are independent, meaning the off-diagonal elements are zero, R is equal to the covariance of the observation (σ_d^2). The weight of model realisation i can be written as:

$$w_i = P(y_i|d) = \frac{P(d|\, y)}{\sum_{j=1}^{n} P(d|\, y_j)} = \frac{\exp\left(-\frac{(d-y_i)^T(d-y_i)}{2\,R}\right)}{\sum_{j=1}^{n} \exp\left(-\frac{(d-y_j)^T(d-y_j)}{2\,R}\right)} \tag{6.3}$$

So the posterior probability is giving the weight (w_i) of the model realisation (particle) i, and the total should be equal to 1 (see Figure 6-14).

When the weight of each particle is calculated, the particle with the highest weight is selected, cloned (sampled into several particles) and the forward calculation is carried out. The Residual Resampling is carried out by firstly deciding about the sample size N, depending on the available computational power. Then the weight is multiplied by the sample size, obtaining $k_i = N.w_i$. The particles with k_i larger than 1 are selected and cloned based on rounding down of k_i to an integer value p_i. The residuals of k_i, which are obtained from the original values deducted by the closest integers, are used to resample the remaining particles $(N - \sum_{j=1}^{N} p_i)$ to reach a sample size of N. The residual weight is calculated by:

$$r_i = \frac{w_i.N - p_i}{N - \sum_{j=1}^{N} p_i} \tag{6.4}$$

The remaining particles are taken by uniform sampling from a cumulative distribution function of r_i.

6.5.2 Data assimilation of EO ETa in Rijnland SIMGRO model

The EO ETa maps presented in Section 4.5 are assimilated in the Rijnland SIMGRO model. The eight-day ETa composite map is first separated into daily maps by filtering the maps with the day of the year (DOY) map. The resulting daily maps sometimes contain only a small number of pixels with data, resulting in a limited coverage of the area, especially in winter period. The maps with low area coverage are removed and only maps with more than 30% coverage are used for data assimilation. However, a map can be chosen to be assimilated even though the coverage is below 30% if the pixels are well-spread over the area. The selected maps are then used for DA. In a DA process the hydrological model's inputs, states or outputs are modified such that the simulated ETa becomes closer to the EO data. In this research, we use DA to update model inputs (precipitation and reference evaporation). The time steps of assimilation correspond to the dates of the selected ETa maps. After data assimilation the simulated discharge is compared with observed discharge and with simulated discharge without DA. This process is represented in Figure 6-15.

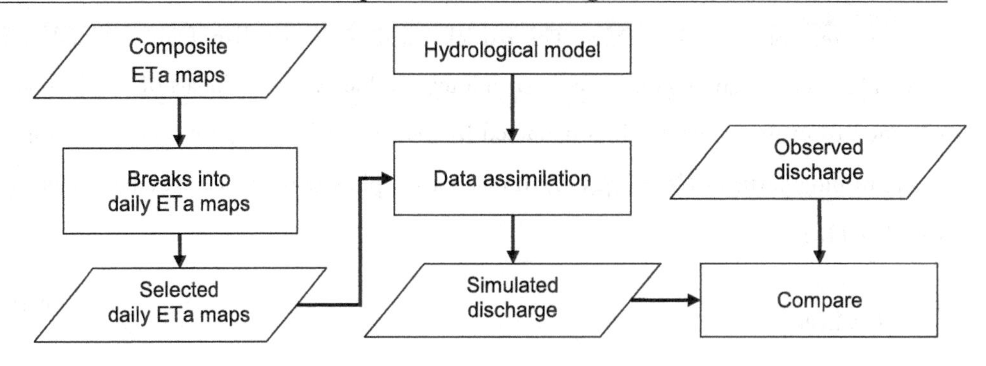

Figure 6-15. Overview of modelling process with data assimilation of EO ETa maps

Software and implementation of Data Assimilation Framework

The PCRaster Python Framework with the PCRaster Python extension is able to handle spatially distributed hydrological models and process these with Particle Filters (Karssenberg and Jong 2005; Karssenberg et al. 2010). The framework offers a combination interface for model construction and optimization. Further information can be found at the PCRaster website (http://pcraster.geo.uu.nl/).

Implementation of the PCRaster Python Framework for this research needed an additional Python script to translate the ETa maps from EO and SIMGRO output into a readable format for PCRaster. Furthermore, the SIMGRO model needed to be called from the framework using system calls. The framework prepares the input, runs the software, and reads the SIMGRO output for the data assimilation process. An illustration of the particle filter framework is presented in Figure 6-16.

The Particle Filter is implemented for a simulation from July 28th to October 11th 2013. This period is selected to include a long dry period followed by a high discharge peak. Due to the high computational demand, for study the sample size is set to 100 particles. To make the resampling process faster, the Rijnland area is divided into five zones, and the particles are weighted based on the averaged ETa in those zones. The five areas are constructed by clustering the model grid cells based on the elevation, soil type, and land use, as seen in Figure 6-17. The Schiphol zone is a very low-lying area with an average elevation of -3.75m+NAP, with agriculture as the main land use. The dunes in the west part consist of sandy soils with shrubs and bare land. The middle Rijnland zone has a large urban area coverage, with an average elevation of 0.20m+NAP. The South Rijnland zone is mainly covered by grassland and agricultural areas with peat soil. The last zone, in the south-west corner, is the Woerden area that discharges separately into the main storage basin of Rijnland as described in Chapter 5.

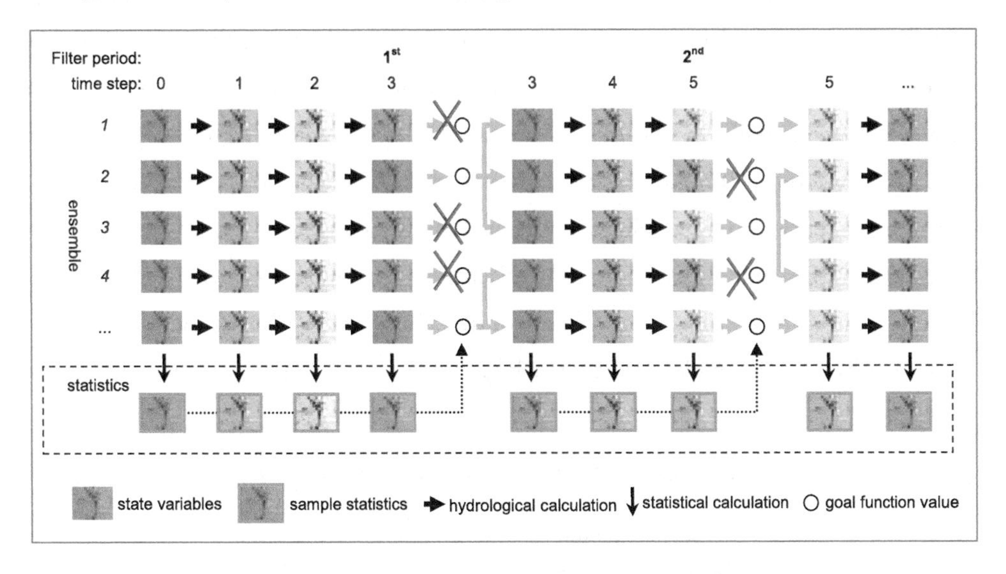

Figure 6-16. Particle Filter application in this study

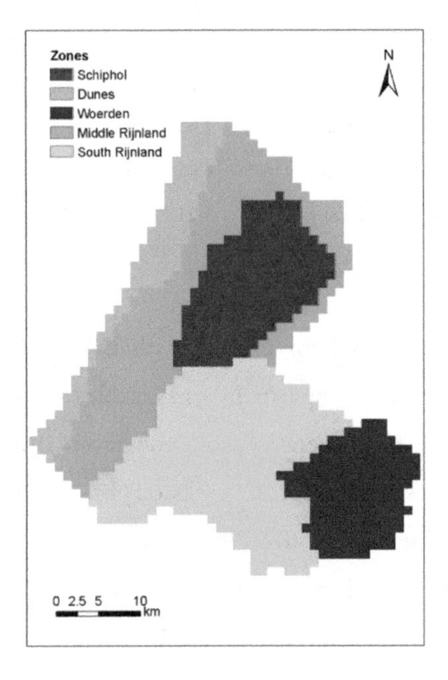

Figure 6-17. Zone map for particle weighting

Resampled variables

Rainfall and reference evapotranspiration have a large influence on the calculation of discharge and actual evapotranspiration, so they are chosen as the input variables that are randomised for residual resampling. The range of the random multiplier (resampling factor) was determined by evaluating model results of 2011 using different multiplier factors for precipitation and reference evaporation (Table 6-5). The precipitation resampling factor results in a normal distribution of discharge with the mean of 1 and the standard deviation of 0.2, (i.e. the 5th percentile is 0.67, and the 95th percentile is 1.33). With the precipitation resampling factors within this range, the model still gives an acceptable NSE and PBIAS for 2011 discharge simulation as seen in Table 6-5. Following the same consideration, a standard deviation of 0.3 is chosen for the reference evapotranspiration resampling factor which gives 0.5 as the 5th percentile and 1.5 as the 95th percentile.

Table 6-5. Results of factor range analysis for year 2011

Model performance with different precipitation factors

factor	0.50	0.60	0.70	0.80	0.90	1.00	1.10	1.20	1.30	1.40	1.50
NSE	0.29	0.50	0.68	0.81	0.88	0.86	0.76	0.58	0.30	-0.08	-0.55
PBIAS	80.16	66.63	52.17	36.90	21.02	4.64	12.15	29.29	46.73	64.42	82.23

Model performance with different ref. evapotranspiration factors

factor	0.50	0.60	0.70	0.80	0.90	1.00	1.10	1.20	1.30	1.40	1.50
NSE	0.64	0.73	0.80	0.83	0.85	0.86	0.86	0.85	0.84	0.82	0.80
PBIAS	36.53	27.28	18.54	10.31	3.73	4.64	11.37	17.65	23.55	29.20	34.60

6.5.3 Data assimilation results

The EO ETa maps are prepared for the run with DA from 28th of July to 11th of October 2013. There are seven composite maps available during this period. Of the 56 (7 * 8) daily ETa maps that are generated from the seven composite maps, ten are of sufficient quality to be selected for assimilation by the hydrological model. The assimilation time or the filter time is then decided to be on the day of the selected maps (the Rijnland SIMGRO model is run with a daily time step). The chosen assimilation times are 1, 5, 14, 19, 23, and 26 of August, 5, 6, 15, and 16 of September 2013. Some of the selected ETa maps are presented in Figure 6-18 as examples. The maps of 15th and

16th of September have a coverage area of less than 30%, however the pixels are well-spread over the Rijnland area. The example of 26th of August is an example of the coverage of more than 30% but with the non-uniform spread over the area (the eastern half of the area has no data).

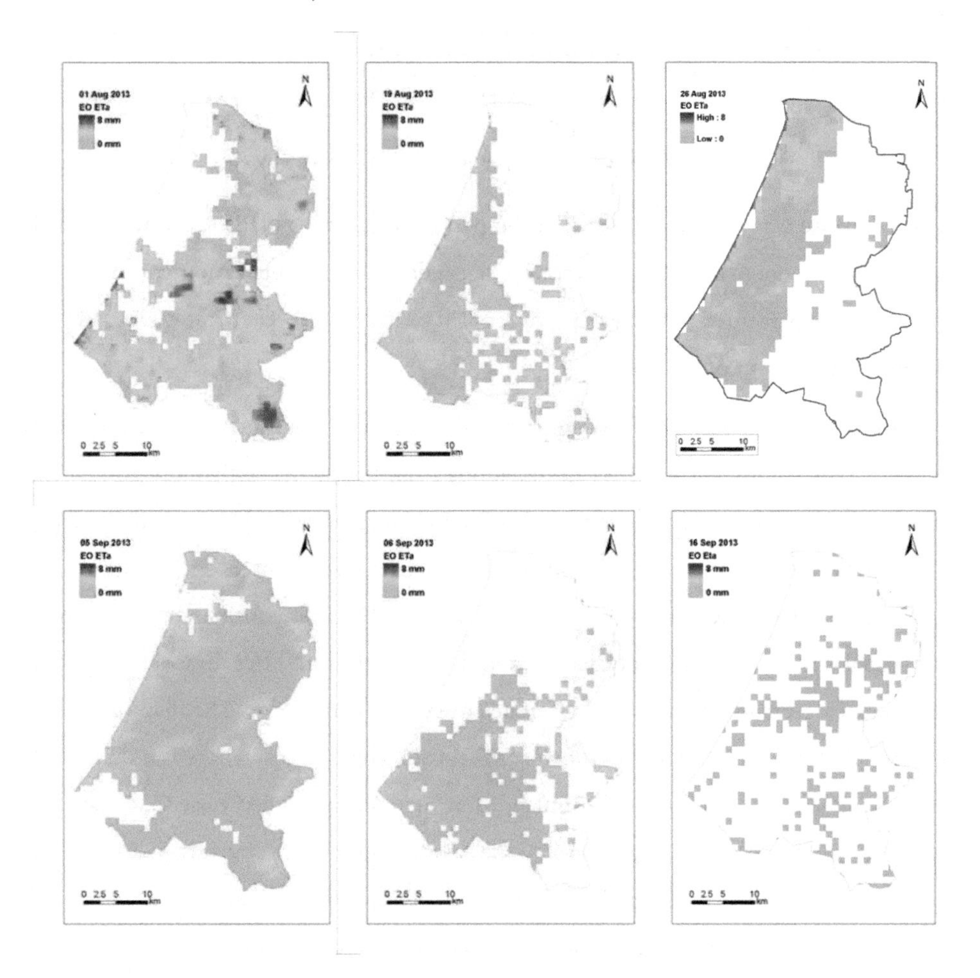

Figure 6-18. Some of the selected EO ETa maps for data assimilation

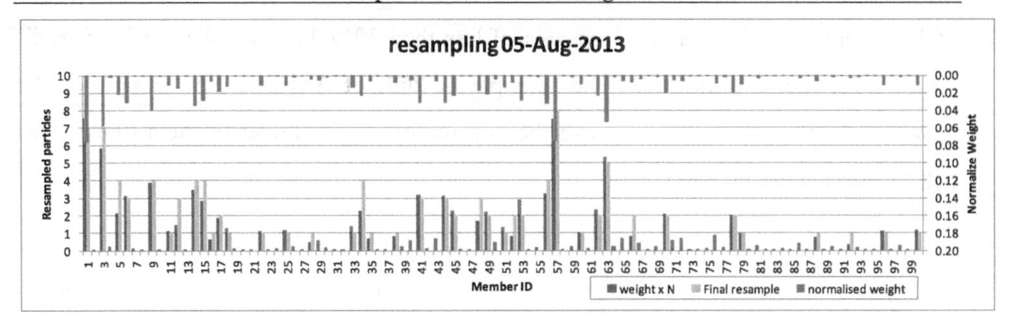

Figure 6-19. The resampling of particles with residual resampling

Particle collapse was not observed when the particle resampling statistics were analysed. For example, Figure 6-19 shows the resampling statistics for the filter moment of August 5 2013. For the presented run, the most resampled particle is particle 57; it was resampled into eight particles out of the 100.

The simulation run with DA is compared with the run without DA for ETa and discharge. Spatial comparison of ETa estimations from the model, with and without DA, is presented in Figure 6-20. The spatial distribution of ETa varies although the average ETa is the same. The ETa comparison for August 26th for example, shows almost the same average ETa for Rijnland, both with and without DA, but considerable differences in its spatial distribution: the simulated ETa in the grid cells can differ up to 1.7 mm/day, while the area-average ETa shows a difference of only 0.02 mm/day. The map of August 1st shows that the DA resulted in a lower ETa in the coastal area, where the dunes are, and the higher ETa for the rest of Rijnland. On the contrary, when observing the ETa difference on August 19th (Figure 6-20) the model with DA simulates a lower ETa compared to the model without DA in almost the entire Rijnland area. On August 26th both models show similar ETa, except for small parts in northern Rijnland.

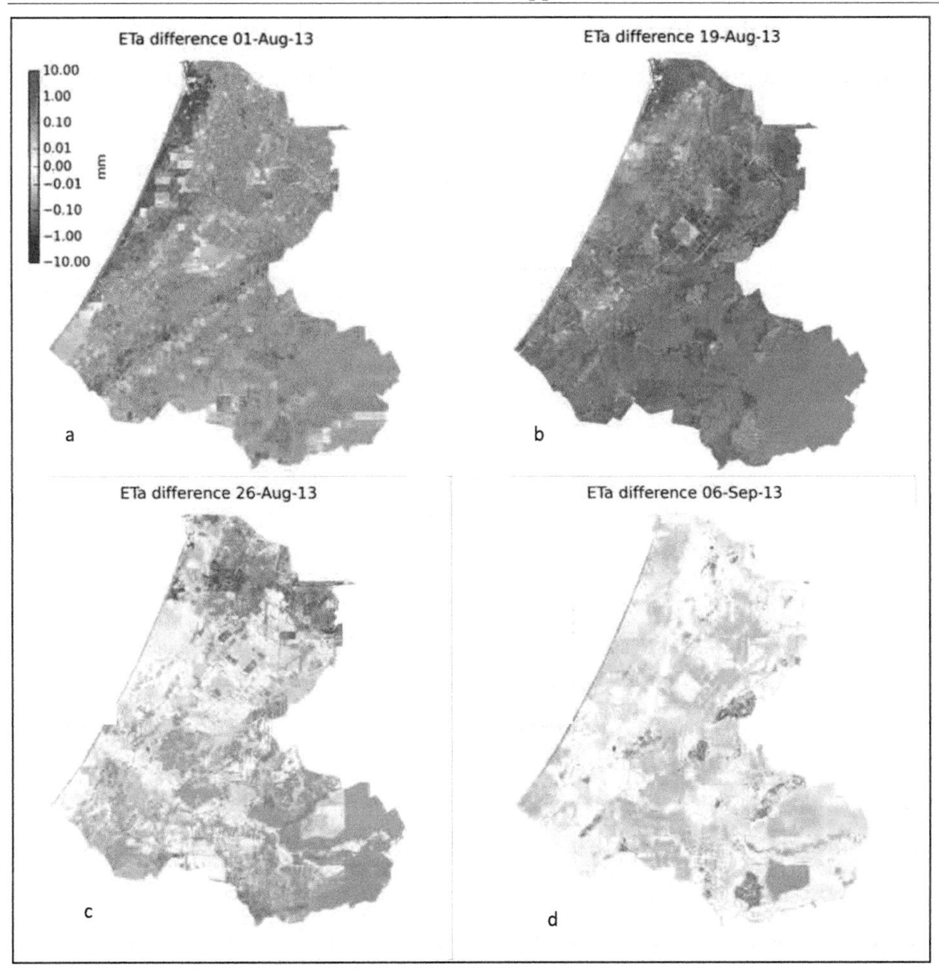

Figure 6-20. Examples of ETa difference maps between simulation with DA and without DA in mm/day, for 1 August, 19 August, 26 August, and 6 September. (A positive difference means that the simulation with DA resulted in a higher ETa)

The ETa differences summed for the entire simulation period of 76 days can be seen in Figure 6-21. The effect of DA can be seen throughout the Rijnland area. For the dune areas in the western part, negative values are observed indicating the calculated ETa with DA gives lower ETa estimation, whereas in the southern part of Rijnland and Woerden area the positive values are obtained. The model with DA also shows higher ETa for grass land, except in some areas where the ground surface elevation is low. Likewise, the model with DA simulates a mixture of higher and lower ETa for agricultural area, but mostly produces a higher ETa compared to the model without DA. Figure 6-20 and

Figure 6-21 show that DA changes the ETa simulation of the model, but there are no ground measurements to further validate the spatially distributed simulated ETa.

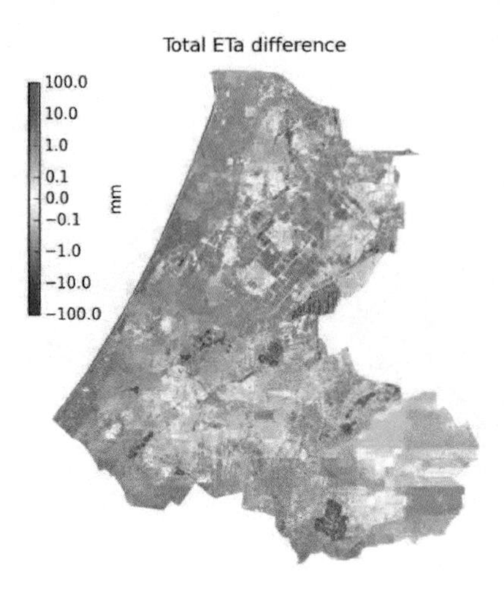

Figure 6-21. Total difference of ETa over the simulation period of 76 days. (A positive difference means that the simulation with DA resulted in higher ETa)

The result of the simulated discharge at the main outlets is presented in Figure 6-22. Negative discharge means that the system has excess water that is being pumped out, while positive discharge means that water is entering the system. Differences are small, and the most notable effect of the DA can be seen in the discharge simulation on September 15 and 16. The model with DA simulates values closer to the observed discharge compared to the model without DA. The measured peak of the high discharge event on September 11 is matched by the simulation with DA, while without DA there is a slight overestimation. The rising limb of the event is simulated too late by both models.

Figure 6-22 also shows that the human influence may affect the observed discharge time series, i.e. from August 1st to 20th, even though the time series is already a three-day moving average. While the model simulation shows a relatively smooth line, the observed discharge line shows small peaks and short low-flow periods. Another indication of human influence can be seen just before the discharge peak on the September 10th. On that day the observed discharge is rising up earlier than the modelled discharge, signalling that pre-pumping might have been applied by the operators. However, this phenomenon may also be caused by an inaccuracy of the rainfall data.

Figure 6-23 shows the cumulative total discharge of the Rijnland area throughout the simulation period. In cumulative discharge time series the effects of human influence (manual operation) are filtered out and the long term processes and water balance simulation can be validated. In the figure, the total discharge volume obtained from the model with DA is closer to the observed data than from the model without DA, with the bias reduced from 14% to 4%. The improvement can be seen in the dry period from August 8th to September 9th. In that period, the simulation result of the model without DA overestimates the discharge, while the model with DA performs well. The improvement can be seen especially in the dry periods where ETa influences to the discharge are relatively high.

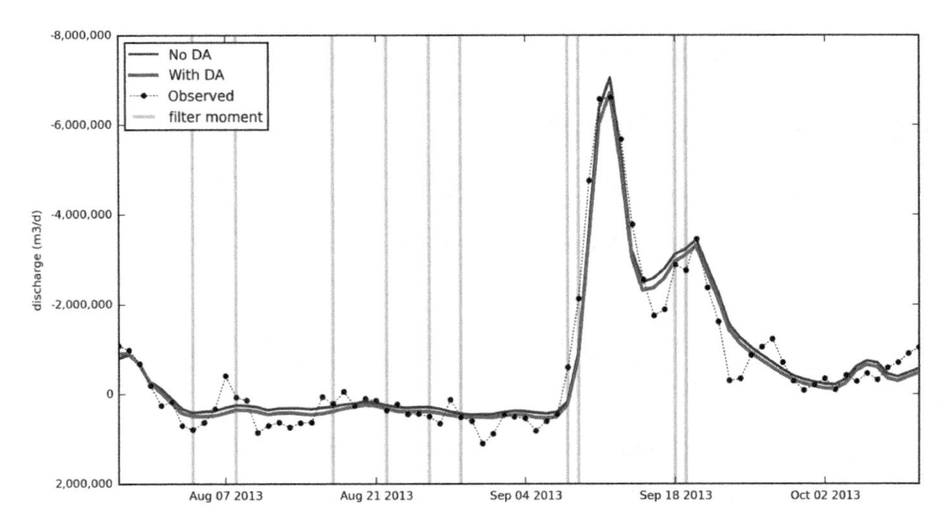

Figure 6-22. Total discharge comparison in the main outlets for a 3-day moving average. The negative discharge indicates that water flows out of the system.

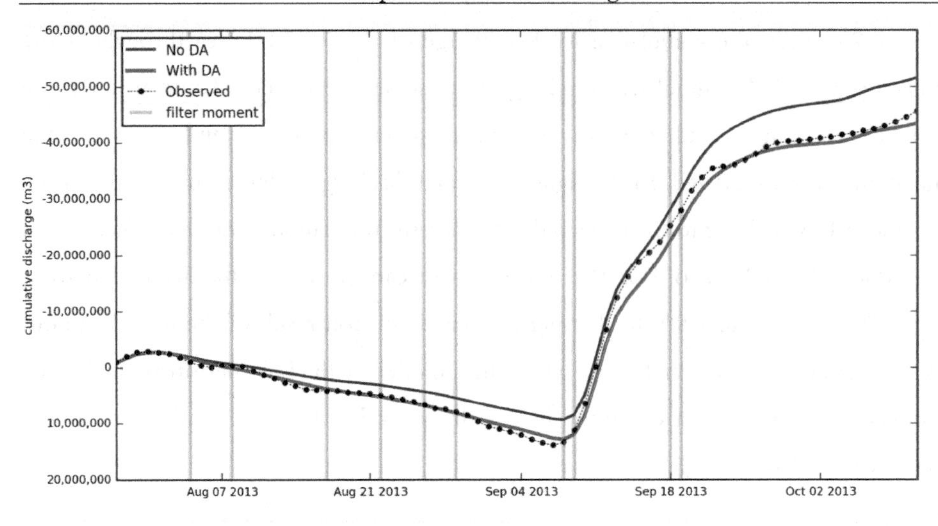

Figure 6-23. Cumulative discharge of the main outlets

6.5.4 Summary and discussion

The results of assimilating EO ETa data in the SIMGRO model of Rijnland showed an increased accuracy of simulated discharge. This confirms that DA is a valuable method in the overall data-model integration framework. The use of an EO ETa composite map causes a reduction in data available for assimilation, however, high quality data is ensured such that the DA can perform at its best.

To limit the computational time, the PF algorithm was set up to use a relatively small pool size of 100 particles, but it should be enough for the considered case study. In case of a lower resolution of the model grid (at the expense of capturing some of the local level drainage processes in the catchment) an ensemble with more particles may be used.

Also, increasing the number of spatial zones for ETa assimilation will enhance the representation of spatial variability. The zones for assimilation can be refined to a pixel by pixel assimilation, in which case the full spatial distribution of the EO data would be taken into account for each pixel. However, increasing the number of zones will result in higher computational time.

Spatially distributed resampling of the parameters can also be done to increase the spatial properties of the DA; this however will require significantly more particles to have a good parameter distribution for each grid cell.

Chapter 7. Multi-model ensemble

In this chapter, each of the available data sources presented in Chapter 5 is used as input to the Rijnland SIMGRO model to form an ensemble of discharge simulations that represents the combined strengths and weaknesses of the input data. This ensemble simulation is analysed against the measured discharge. Discharge simulations from the base model, individual ensemble members, and simple and weighted averages of the ensemble members are compared. Apart from the performance criteria for a deterministic, single, simulation, ensemble simulation performance is analysed with the Brier score and relative operating characteristics.

7.1 Introduction

Following the data-model integration framework presented in Chapter 3 (Figure 3-1), in this chapter, a multi-model ensemble simulation is generated by using the multiple data sources for land use and soil type (catchment characteristics) and for rainfall and evapotranspiration (hydrometeorological data) described in Chapter 4, as input in the SIMGRO model of Rijnland (Chapter 5). The ensemble members are specified in the next section, after which the ensemble simulation results are presented. The last section of this chapter presents the weighting experiments done, and their results.

7.2 Constructing the multi-model ensemble

The SIMGRO base model, with the Thiessen polygon rainfall, European database soil map and LGN land-use map is the first member of the ensemble. The other members are generated by substituting the soil or land use map, precipitation or evapotranspiration input data, with one of the alternative data sources described in Chapter 5. Each unique combination of the base model and data sources forms a new member in the ensemble. Other parameters in the base model are kept constant. Having the two precipitation data sources, three land use maps, two soil maps, and two sources for ETa data, allows for creating an ensemble of 24 members as defined in Table 7-1. The ensemble simulation covers the period from October 2012 to October 2013.

7.3 Ensemble simulation results

7.3.1Performance individual members and ensemble mean

Sample periods of the three-day moving average ensemble discharge simulation are presented together with measured discharge in Figure 7-1. Negative discharge indicates that water is pumped out of the system. The results show that in July 2013 the measured discharge generally falls within the ensemble spread, and the ensemble mean is generally closer to the measured discharge than the base model (LGN_EDB_THI; second member in Table 7-1). In other periods, however, such as in December 2012, the measured discharge is outside the range of the ensemble simulation. Using different data sources covers only part of the data and parameter uncertainty, next to which, there is also model structure uncertainty, and uncertainty in the output itself (discharge observations).

Table 7-1. Ensemble members built from the combination of data sources

	Ensemble member	LULC map	Soil Map	Precipitation	Evaporation
1	LGN_EDB_RDR	LGN	European DB	Radar	Point reference ET
2	LGN_EDB_THI (base)	LGN	European DB	Rain	Point reference ET
3	LGN_DDB_RDR	LGN	Dutch DB	Radar	Point reference ET
4	LGN_DDB_THI	LGN	Dutch DB	Rain	Point reference ET
5	LS5_EDB_RDR	Landsat 5 TM	European DB	Radar	Point reference ET
6	LS5_EDB_THI	Landsat 5 TM	European DB	Rain	Point reference ET
7	LS5_DDB_RDR	Landsat 5 TM	Dutch DB	Radar	Point reference ET
8	LS5_DDB_THI	Landsat 5 TM	Dutch DB	Rain	Point reference ET
9	GBC_EDB_RDR	GLOBCOVER	European DB	Radar	Point reference ET
10	GBC_EDB_THI	GLOBCOVER	European DB	Rain	Point reference ET
11	GBC_DDB_RDR	GLOBCOVER	Dutch DB	Radar	Point reference ET
12	GBC_DDB_THI	GLOBCOVER	Dutch DB	Rain	Point reference ET
13	LGN_EDB_RDR_ET	LGN	European DB	Radar	EO ETa
14	LGN_EDB_THI_ET	LGN	European DB	Rain	EO ETa
15	LGN_DDB_RDR_ET	LGN	Dutch DB	Radar	EO ETa
16	LGN_DDB_THI_ET	LGN	Dutch DB	Rain	EO ETa
17	LS5_EDB_RDR_ET	Landsat 5 TM	European DB	Radar	EO ETa
18	LS5_EDB_THI_ET	Landsat 5 TM	European DB	Rain	EO ETa
19	LS5_DDB_RDR_ET	Landsat 5 TM	Dutch DB	Radar	EO ETa
20	LS5_DDB_THI_ET	Landsat 5 TM	Dutch DB	Rain	EO ETa
21	GBC_EDB_RDR_ET	GLOBCOVER	European DB	Radar	EO ETa
22	GBC_EDB_THI_ET	GLOBCOVER	European DB	Rain	EO ETa
23	GBC_DDB_RDR_ET	GLOBCOVER	Dutch DB	Radar	EO ETa
24	GBC_DDB_THI_ET	GLOBCOVER	Dutch DB	Rain	EO ETa

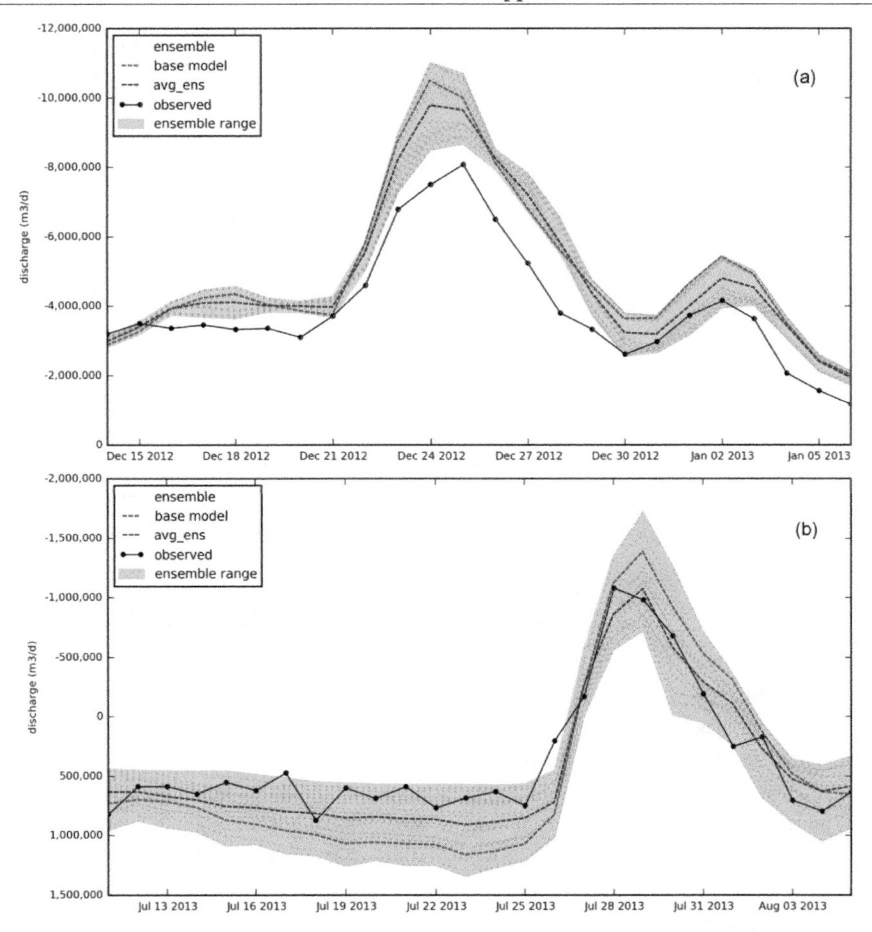

Figure 7-1. Graph of three-day moving average discharge comparison of the ensemble members and observations, for the selected events in winter (a) and summer (b)

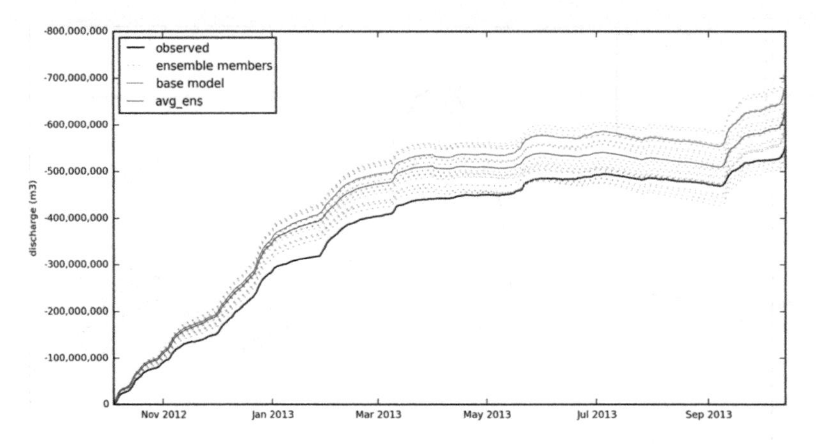

Figure 7-2. Cumulative discharge of the ensemble members

Table 7-2. Model performance for three-day moving average discharge (m3/day)

Ensemble member	Oct 2012 to Oct 2013			Summer			Winter		
	PBIAS	NSE	r	PBIAS	NSE	r	PBIAS	NSE	r
LGN_EDB_RDR	-6.30	0.84	0.95	6.78	0.77	0.95	-8.75	0.81	0.93
LGN_EDB_THI	**-22.18**	**0.81**	**0.96**	**-23.99**	**0.74**	**0.96**	**-21.89**	**0.76**	**0.95**
LGN_DDB_RDR	-7.16	0.84	0.95	0.06	0.79	0.95	-8.54	0.79	0.93
LGN_DDB_THI	-22.86	0.81	0.96	-29.59	0.77	0.96	-21.67	0.74	0.95
LS5_EDB_RDR	3.34	0.86	0.95	56.91	0.78	0.95	-6.36	0.83	0.94
LS5_EDB_THI	-12.92	0.84	0.96	26.07	0.80	0.96	-19.94	0.78	0.95
LS5_DDB_RDR	1.89	0.86	0.95	47.34	0.81	0.94	-6.32	0.81	0.93
LS5_DDB_THI	-14.23	0.84	0.96	17.65	0.84	0.96	-19.94	0.76	0.95
GBC_EDB_RDR	0.30	0.88	0.95	27.93	0.84	0.95	-4.74	0.84	0.93
GBC_EDB_THI	-15.84	0.86	0.96	-4.01	0.84	0.97	-17.99	0.80	0.95
GBC_DDB_RDR	-0.83	0.88	0.95	20.63	0.86	0.95	-4.75	0.82	0.93
GBC_DDB_THI	-16.90	0.86	0.97	-10.58	0.86	0.97	-18.03	0.79	0.95
LGN_EDB_RDR_ET	-13.16	0.83	0.95	-4.96	0.75	0.93	-14.69	0.78	0.93
LGN_EDB_THI_ET	-29.35	0.77	0.96	-38.22	0.73	0.96	-27.74	0.69	0.95
LGN_DDB_RDR_ET	-12.19	0.83	0.94	-8.58	0.76	0.93	-12.88	0.79	0.93
LGN_DDB_THI_ET	-28.19	0.78	0.96	-40.35	0.74	0.95	-25.97	0.69	0.95
LS5_EDB_RDR_ET	-6.35	0.82	0.94	36.35	0.75	0.92	-14.04	0.77	0.92
LS5_EDB_THI_ET	-22.91	0.79	0.96	2.40	0.78	0.95	-27.41	0.69	0.95
LS5_DDB_RDR_ET	-5.59	0.83	0.94	30.76	0.76	0.92	-12.12	0.78	0.92
LS5_DDB_THI_ET	-21.92	0.80	0.96	-1.65	0.80	0.95	-25.50	0.70	0.94
GBC_EDB_RDR_ET	-8.78	0.85	0.95	12.77	0.81	0.93	-12.67	0.79	0.93
GBC_EDB_THI_ET	-20.03	0.81	0.96	14.74	0.79	0.95	-26.24	0.72	0.95
GBC_DDB_RDR_ET	-7.92	0.85	0.94	8.55	0.81	0.93	-10.90	0.80	0.92
GBC_DDB_THI_ET	-23.94	0.82	0.96	-24.08	0.83	0.96	-23.87	0.73	0.95
avg_ens	**-13.08**	**0.85**	**0.96**	**5.12**	**0.82**	**0.95**	**-16.37**	**0.80**	**0.94**

The cumulative discharge graph in Figure 7-2 shows that the ensemble mean (avg_ens) gives a closer estimation of the observed discharge than the base model. Cumulative discharge evaluates the long term water balance and exposes any seasonal errors that occur. A high deviation of the ensemble mean from the observed cumulative discharge occurred from November to December 2012. For the remainder of the simulation period the ensemble mean and observed cumulative discharge run in parallel. The base model overestimates the discharge.

The performance of the different model set-ups, the members, for three-day moving average streamflow can be seen in Table 7-2. All ensemble members show a very good model performance in terms of NSE, according to performance rating by Moriasi and Arnold (2007). However, two of the models have an unsatisfactory PBIAS of more than 25%, which are LGN_EDB_THI_ET and LGN_DDB_THI_ET (members 14 and 16 in Table 7-1), showing that the combination of LGN, THI and ET results in low performance. Furthermore, Furthermore Table 7-2 shows that in general the model performance is lower in the summer period compared to the winter period. The difficulties to model the discharge in summer in Rijnland may be partly due to the difficulty to model the inlets and flushing that are applied during this period.

The ensemble mean performs well, with high NSE, even when divided into summer and winter period. Although some of the individual ensemble members perform better on some of the performance criteria over a certain period (year or season), overall the ensemble mean's performance is higher. The comparison of performance of the base model and the ensemble mean is presented in Table 7-3. In particular, the PBIAS is reduced when applying the multi-model ensemble. Note that for the base model the PBIAS over the experimental period of October 2012 to October 2013, is much higher than for the validation period of 2011 (Chapter 5).

Table 7-3. Model performance comparison between the base model and the ensemble mean

Performance rating	LGN_EDB_THI (base model)	Ensemble mean
PBIAS	-22.18	-13.08
NSE	0.81	0.85
r	0.96	0.96

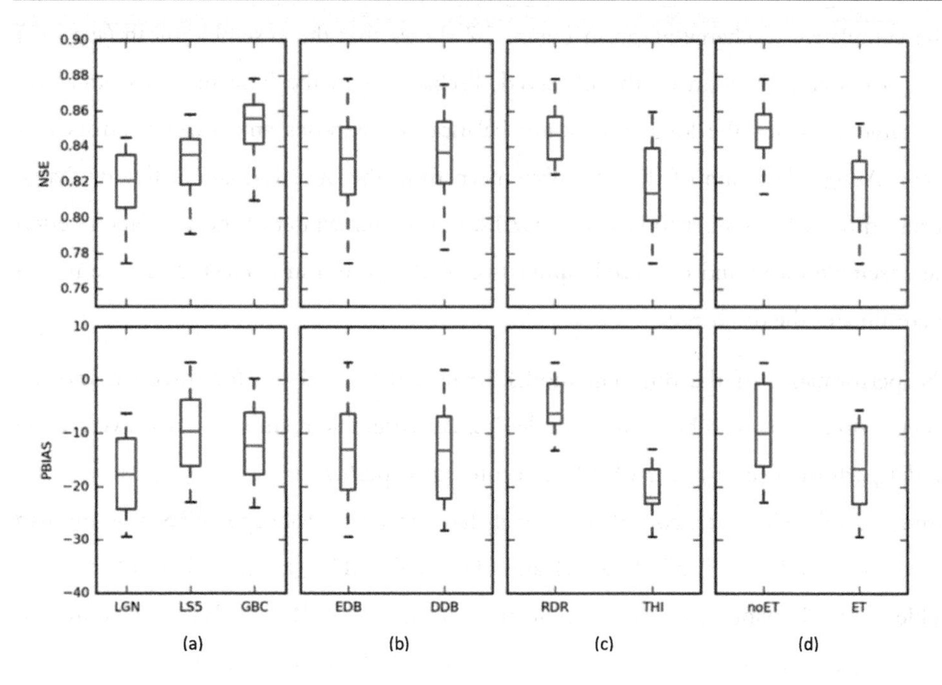

Figure 7-3. Box plot of the model performances for the simulation period October 2012 to October 2013, clasified by data source: land-use (a), soil type (b), rainfall (c), and ET (d)

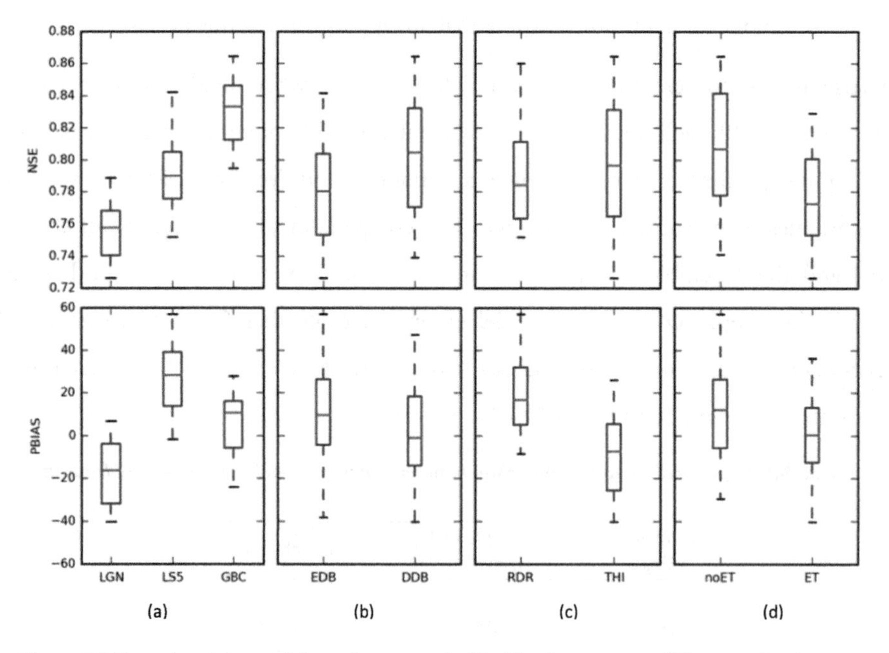

Figure 7-4. Box plot of the models performance clasified by data sources differences for the summer period, land-use (a), soil type (b), rainfall (c), and ET (d)

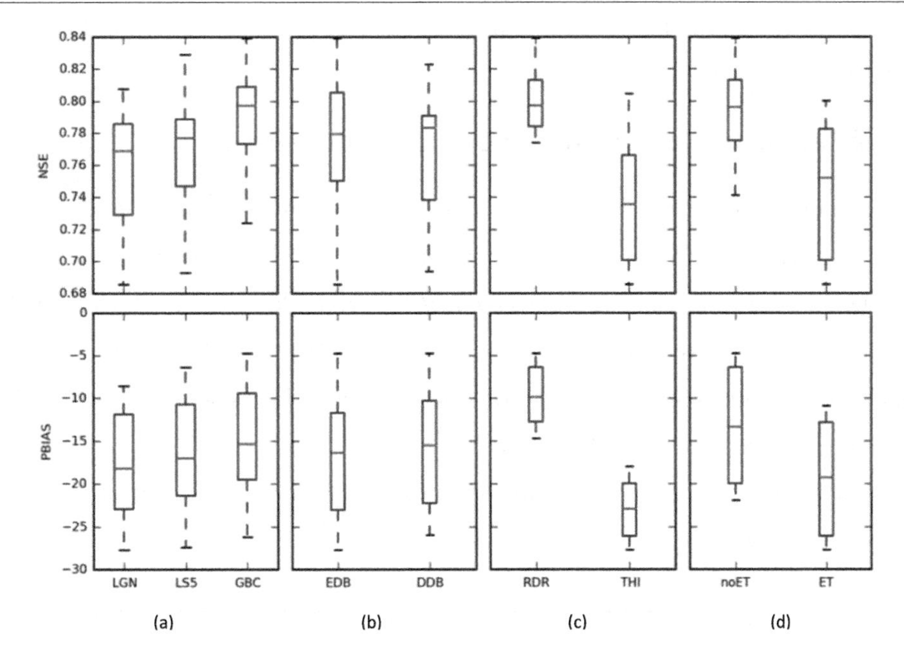

Figure 7-5. Box plot of the models performance clasified by data sources differences for the winter period, land-use (a), soil type (b), rainfall (c), and ET (d)

Figure 7-3 presents the influence of each data source on model performance. The soil type has the least influence on model performance for the simulation period, as seen in Figure 7-3b. Figure 7-3a shows that the LGN land use resulted in the lowest model performances, compared to the other two land use maps, both for NSE and PBIAS. The models with the GBC land use resulted in higher NSE compared to the model with the LGN and the LS5 land use map. However, the models using LS5 have higher performance when looking at PBIAS. Figure 7-3c shows that the rainfall data have the highest influence on the model performance, with the radar rainfall resulting in higher model performances compared to the rain gauge in terms of NSE and PBIAS. Figure 7-3d show that the models with EO ETa have a lower performance than the models without EO ETa, (i.e. using ground station reference evapotranspiration).

However, when analysed for the summer period only, when results are less dominated by rainfall, and ET is relatively more important, the models with EO ETa generally have lower PBIAS compared to the models with ETr. The soil maps also have a higher influence in summer, with the models with DDB data showing better performance than the models with EDB. The summer and winter period graphs can be seen in Figure 7-4 and Figure 7-5.

7.3.2 Performance of the ensemble simulation

Brier scores (BS) for the simulation period are presented in Figure 7-6. Increasing discharge thresholds (Q_0) are selected from 500,000 m^3/day to 5,500,000 m^3/day, with a 500,000 m^3/day interval. The 4,500,000 m^3/day threshold corresponds to the 90th percentile of the observed discharge (4,350,000 m^3/day). The BS of the ensemble simulation is plotted together with the BS of each individual member listed in Table 7-1, and the ensemble mean. For the BS of the individual members and ensemble mean, a 100% probability is associated with the simulated value, as if no uncertainty is associated with these simulations.

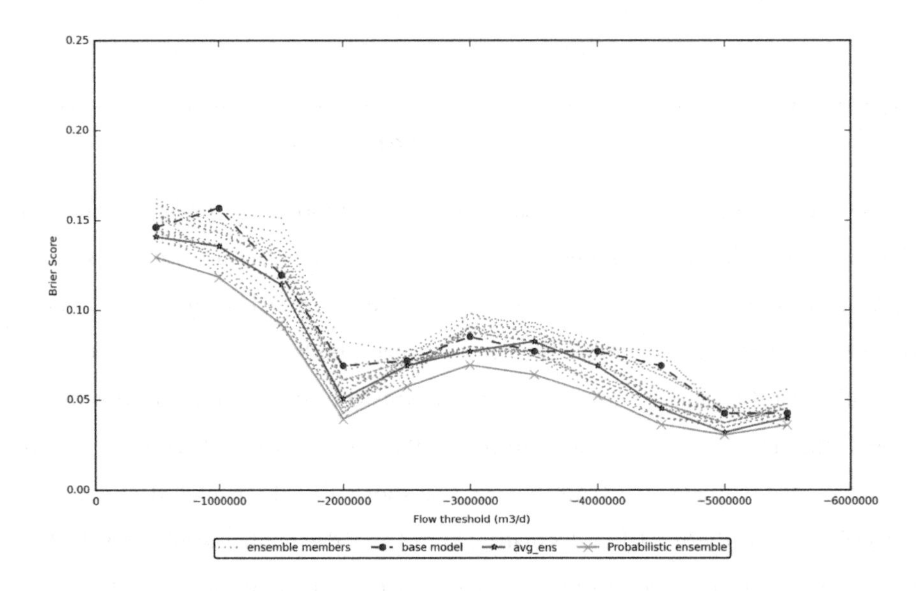

Figure 7-6. Brier score for the simulation period of October 2012 to October 2013 (lower is better)

Under that assumption, the probabilities of event threshold exceedance simulated by the ensemble, are equally or more accurate (lower is better) than for any of the deterministic simulations, for all event thresholds. The ensemble mean (avg_ens) results in a lower BS than most of the individual ensemble members for most of the thresholds. For the 5,000,000 m^3/day threshold, the ensemble mean has a lower BS compared to all other members. The same pattern also shows when the time series are separated for summer and winter periods.

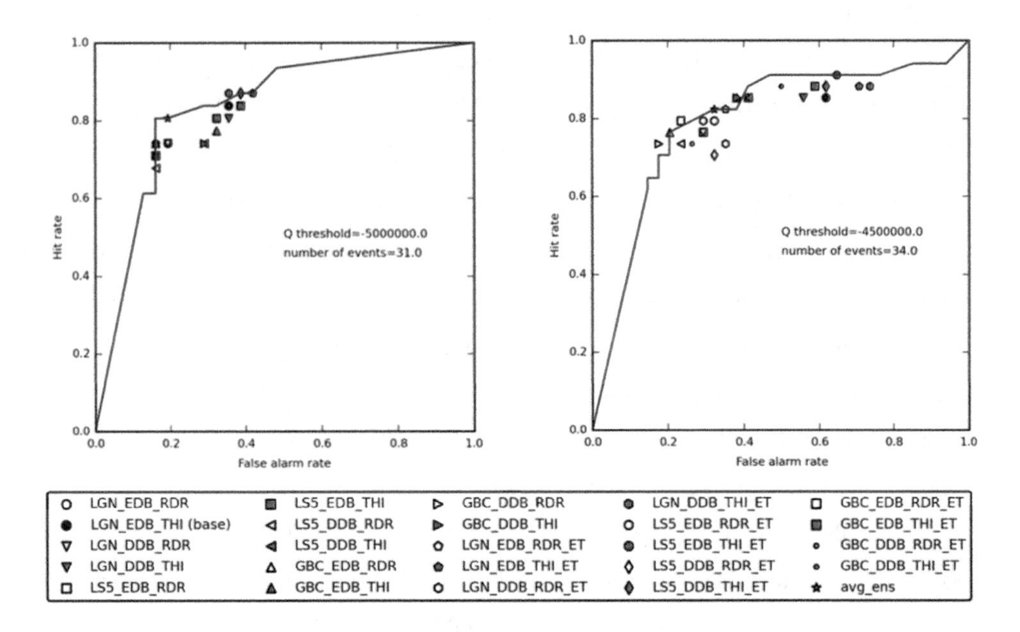

Figure 7-7. ROC diagram for the simulation period from October 2012 to October 2013

Figure 7-7 shows the Relative operating characteristic (ROC) diagram of the ensemble (blue line). The ROC curve is the line that connects various levels of probability as given by ensemble simulation. The larger the ROC area, which is the area under the curve, the more skilful the simulation is. Because the diagonal line indicates no skill, a good model ensemble should give ROC area higher than 0.50. From the figure it can be seen that the ensemble probability simulation gives a good result. The ROC area of the ensemble simulation is 0.83 for the flow threshold of 5,000,000 m^3/day, and 0.80 for 4,500,000 m^3/day. Table 7-4 presents the ROC curve areas for the other thresholds. With the ROC higher than 0.50 for all thresholds, it can be concluded that the ensemble has a good performance.

Figure 7-7 also plots the HR-FAR position of the deterministic simulations (assuming 100% probability) from the individual ensemble members from Table 7-1, and the ensemble mean. All ensemble members are showing good skill in simulating exceedance of the discharge threshold. For deterministic simulations, the ROC area is calculated by the trapezoid area between the coordinate of HR-FAR position, upper right corner (1,1), lower right corner (1,0), and lower left corner (0,0). From the calculations with 5,000,000 m^3/day (31 observed events), the ensemble mean gives a ROC area of 0.81 with a FAR of 0.19 and a HR of 0.81, while the base model has a ROC area of 0.67 with a FAR of 0.35 and a HR of 0.84. Although the ensemble mean has a slightly lower HR, it has a significantly higher ROC area due to the lower FAR. The ROC area of the ensemble mean is higher than for all other members, further strengthening that a simple average of a multi-model ensemble, can already be better than a single model. Note, however, that even though the ROC area of the ensemble mean is the highest, there are six individual members (models) with a higher HR, but having a higher FAR. This information can be given to the water managers to let them decide what to use, and how. It could be, for example, that for the Rijnland water system a false alarm does not have much negative impact, and water managers may give priority to a high hit rate. Also, the results may differ per threshold. For the threshold of 4,500,000 m^3/day (34 observed events), for example, three members have a slightly higher ROC area than the ensemble mean. However, these three members have a considerably lower HR.

Table 7-4. Comparisons of ROC area performance result

Threshold (m^3/day)	Number of observed events	LGN_EDB_THI (base)	ROC area avg_ens	ROC area ensemble
2,000,000	106	0.88	0.91	0.95
3,000,000	66	0.76	0.78	0.82
4,000,000	44	0.67	0.71	0.81
4,500,000	34	0.62	0.75	0.80
5,000,000	31	0.74	0.81	0.83

The ensemble as a whole has a higher ROC area still, compared to the ensemble mean and the individual ensemble members (Table 7-4). Note, however, that these are not fully comparable as the ensemble as a whole is represented by a curve instead of a point. In the probabilistic ensemble, the decision maker can choose for which simulated probability of exceedance to issue an alarm. The decision maker can issue an alarm when only one member of the ensemble reports an event, or specify that, for example, at least 50% of the ensemble members should simulate that the threshold is reached. Again, this leaves the freedom with the water manager to decide on what information to use, and how.

7.4 Weighting of ensemble members

In the previous sections the members were considered to have equal weights, and the ensemble mean was calculated using simple averaging. In the following sections static and dynamic weighting schemes are tested to analyse whether the performance of the ensemble simulation can be further improved.

7.4.1Static weighting

Static performance-based weighting is applied. It is based on 2012 performance as shown in Table 7-5. The weighting scheme uses the normalized inverse distance, in which the ensemble members are weighted based on the inverse distance to the perfect performance rating in 2012 (Section 3.4). For example, the inverse distance weighting based on PBIAS is calculated by the inverse distance to perfect rating of PBIAS, which is 0. For weighting based on a single performance metric PBIAS is selected, because there is high variability in the PBIAS score between the ensemble members (Table 7-2). All performance criteria can also be combined into one weight, using Euclidian distance for more than one dimension (in this case four dimensions: PBIAS, RSR, NSE and r). The inverse distance is then normalized to 1.

The weights found based on the performance of the ensemble members in 2012, are subsequently applied to the period from January to October 2013. The weighted ensemble mean performance is analysed and compared to the performance of the simple average ensemble mean as shown in Table 7-6. It can be seen that the improvement is not very high. The weighting based on the PBIAS gives a small improvement only for PBIAS score itself, not for the other metrics.

Table 7-5. The ensemble member PBIAS performance in January to October 2012 and the related weight

members	Performance	weight
	PBIAS	w_PBIAS
LGN_EDB_RDR	-5.961	0.0611
LGN_EDB_THI	-17.215	0.0212
LGN_DDB_RDR	-4.084	0.0892
LGN_DDB_THI	-15.330	0.0238
LS5_EDB_RDR	1.678	0.2171
LS5_EDB_THI	-10.204	0.0357
LS5_DDB_RDR	3.087	0.1180
LS5_DDB_THI	-8.823	0.0413
GBC_EDB_RDR	1.884	0.1933
GBC_EDB_THI	-9.513	0.0383
GBC_DDB_RDR	3.101	0.1175
GBC_DDB_THI	-8.354	0.0436

Table 7-6. Performance of weighted time series for 2013, the last column is weighted to all four performance criteria

Performance of the	Weighted to		
ensemble mean	equal	PBIAS	PBIAS.RSR.NSE.r
RSR	0.515	0.520	0.514
PBIAS	-4.396	3.602	-4.202
NSE	0.735	0.730	0.736
r	0.904	0.895	0.904

7.4.2 Dynamic weighting

In dynamic weighting, the weights of individual members change in time. In this section, the weight is determined by errors of previous time steps or trend of the time series in the previous time steps, and applied in the weighting formula of equation 3-2. The time frame to determine the errors and trend can be one time step or several. The errors and trends are also utilised to dynamically filter out the least performing members (assigning zero-weight), either by ranking the members based on the error, or based on the trend that the majority of members indicates (increasing or decreasing discharge in the previous time steps).

Table 7-7. Performance of ensemble mean for different dynamic weighting schemes

Performance metric	avg_ens	Majority trend filter	Error 1d	Error 2 day	Error 3 day	Error 1 day 50% filter	Error 2 day 50% filter	Error 3 day 50% filter
PBIAS	-13.08	-12.90	-10.19	-11.09	-11.42	-8.57	-9.78	-10.23
RSR	0.38	0.39	0.36	0.36	0.37	0.32	0.33	0.34
NSE	0.85	0.85	0.87	0.87	0.86	0.90	0.89	0.88
r	0.96	0.96	0.96	0.96	0.96	0.97	0.96	0.96

Table 7-7 presents the performance metrics of the ensemble mean time series resulting from the different dynamic weighting schemes. The table shows that the majority trend filter did not give a significant improvement. The bias score was improved slightly but no improvements are seen for the other performance metrics. The small improvement may be due to a lack of opposite trends between simulated and observed discharge for the significant events. The weighting based on yesterday's error has given a noticeable improvement in PBIAS, NSE and RSR score. Increasing the number of time steps back to take into account, however, did not improve the performance further, as seen in the Error 2 day and Error 3 day of Table 7-7. Dynamically giving the 50% poorest performing members a zero weight, further improves the performance, as seen in Table 7-7 with 50% filter. Compared to the simple average (avg_ens) the improvement is significant. PBIAS decreased from 13.08% to 8.57%, and NSE improved from 0.85 to 0.9 for the weights based on yesterday's error.

The impact of the dynamic weighting scheme with filtering on the ensemble simulation performance expressed by ROC area, can be seen in Table 7-8. For 4 of the five event thresholds analysed, the ROC area increased by the dynamic filter weighting.

Table 7-8. ROC area comparison of the base model, unfiltered ensemble and dynamically filtered ensemble

Threshold (m^3/day)	LGN_EDB_THI (base)	ROC area ensemble (unfiltered)	ROC area filtered ensemble
2,000,000	0.877	0.948	0.954
3,000,000	0.758	0.819	0.945
4,000,000	0.670	0.809	0.842
4,500,000	0.618	0.795	0.916
5,000,000	0.742	0.833	0.804

7.5 Summary

The results in this chapter showed that the multi-model ensemble generated from multiple data sources for catchment characteristics (land use and soil type) and hydrometeorological input (rainfall and evapotranspiration), improved the simulation of discharge from the Rijnland water system. The ensemble mean generally performed better than the base model and most other individual members in terms of NSE, PBIAS, and correlation coefficient. ROC analysis showed that the probabilistic ensemble generally performs better than individual members in simulating exceedance of discharge thresholds.

The method was applied with the physically-based spatially distributed SIMGRO model of Rijnland, which was parameterised based on expert judgement, with limited calibration (Chapter 5). Only the base model was calibrated. There was no recalibration after using alternative data sources to construct the other ensemble members. It is, therefore, possible to attempt to calibrate each of the individual ensemble members to try to further improve the simulations' performance.

The uncertainty in observations of the main output variable, measured discharge, is not considered in this research, which may also affect the ensemble and individual member performance results.

The static weighting results showed that the simple equal weighting was already giving good results. The scheme with previous-year performance only gave a small improvement to the ensemble mean as compared to the mean with equal weights. The weighting using combined performance metrics also gave only a small improvement. The dynamic weighting using previous-day error resulted in stronger improvements. Dynamically giving zero weight to half of the members with high previous-day error was resulting in a significant improvement of ensemble mean NSE. Note, however, that this kind of dynamic weighting is relevant for forecasting applications, and not for simulation applications. The results of the different weighting methods are specific to the considered case study of Rijnland. In other case studies, other weighting schemes may yield the best performance results.

Chapter 8. Towards implementation in operational systems[2]

This chapter explores whether the data-model integration framework presented in this thesis, could be applied in today's operational systems for hydrological simulations and real-time forecasts. To accommodate the framework, an operational system needs to be able to retrieve different data formats from multiple sources and feed them to a hydrological model. It should enable scheduling model runs using different data combinations, saving the model outputs, followed by processing, e.g. weighting, and visualisation of results. One such system, briefly discussed as example in this chapter, is the MyWater platform developed by Hidromod *(hidromod.com) for the MyWater project (*Hartanto *et al. 2015).*

For the data-model integration framework developed in this thesis to be used in practice by hydrological modellers and water managers, supporting ICT technologies need to be available to manage the high computational and storage requirements of a multi-model ensemble approach. The MyWater platform has been tested with multiple data sources, multiple hydrological and hydrodynamic models, and for multiple case studies. The MyWater platform was developed by Hidromod as part of the MyWater EU Project (http://aquasafeonline.net/; Hartanto et al. 2015).

The MyWater research project covered five case studies with distinct geographical locations and characteristics; Umbeluzi catchment in southern Africa (Mozambique, Swaziland and South Africa), Tamega catchment (Portugal and Spain), Rijnland area (The

---

[2] This chapter is based on I. M. Hartanto, Carina Almeida, Thomas K. Alexandridis, Melanie Weynants, Gildo Timoteo, Pedro Chambel-Leitão, Antonio M.S. Araujo (2015) Merging earth observation data, weather predictions, in-situ measurements and hydrological models for water information services. Environmental Engineering and Management Journal 14(9)

Netherlands, Section 4.1), Nestos River (Greece and Bulgaria) and Queimados River (Brazil). For each of the five case studies, earth observation (EO) data, such as leaf area index (LAI), actual evapotranspiration (ETa), and soil moisture (SM) were provided. Additionally, the following maps were used to set-up the catchment models: Soil map, DEM, Land Use/Land Cover, and Drainage or River network. Ground station (in-situ) time series data consisting of rainfall, streamflow, temperature (daily min and max), derived potential evapotranspiration, and wind speed, were used as input or for calibration and validation.

Hydrological and hydrodynamic models have been developed for all case studies. Types of models and modelling system software have been selected on the basis of the unique characteristics of each catchment. The aim of having five different case studies, with different modelling systems, and need for different water management services, was to ensure that the MyWater platform would be a flexible information system for water management.

The following modelling systems were used. Mohid Land (Trancoso et al. 2009) is a spatially-distributed hydrological modelling system, chosen as the main model in Tamega, Queimados and Nestos. SWAT (Neitsch et al. 2005), a semi-distributed modelling system, is the main system for the Umbeluzi case study. SIMGRO (van Walsum and Veldhuizen 2011) for the Rijnland area, as presented in this thesis (Chapter 5). In addition to hydrological models, hydrodynamic models, i.e. a PriceXD (Price et al. 2014) overland-flow model and a Mohid Water (Trancoso et al. 2009) 2D/3D hydrodynamic model, drainage models (SWMM (Huber et al. 2005) storm water modelling and management tool), and reservoir models (CEQUALW2 (Cole and Wells 2000)) have been set-up and incorporated in the MyWater platform.

Due to the different data requirements, such as data type and spatial and temporal resolution, not all models can use the same data. On the other hand, with the availability of multiple data sources, the model can sometimes be run repeatedly for each data source, following the ensemble-based integration framework proposed in this thesis (Chapter 3).

The MyWater platform enables integration of the in-situ, EO, meteorological model output data, and the hydrological models presented in the previous paragraphs to provide real-time water management services, such as streamflow forecasts and flood and drought early warning (http://www.hidromod.com/, http://aquasafeonline.net/, Hartanto et al. 2015). A new data source or model can be added to the system by using the available API (application program interface) and converter in the MyWater platform. Day-to-day water management can be supported in a flexible way. Water managers can see the behaviour of the catchment clearly from different points of view. Water managers also can get their needed reports in the way they want these to be, due to the highly customisable workspaces (Figure 8-1) and reports from the platform.

The platform was tested operationally successfully, showing that the data-model integration framework presented in this thesis can be accommodated in today's state-of-the art operational information and decision support systems for water management. The platform is able to retrieve data from different sources and feed them into hydrological models, and then process the results, e.g. weighting, as necessary. The information presentation is capable to visualise multi-model outputs such as multiple time series and multiple maps resulting from an ensemble simulation.

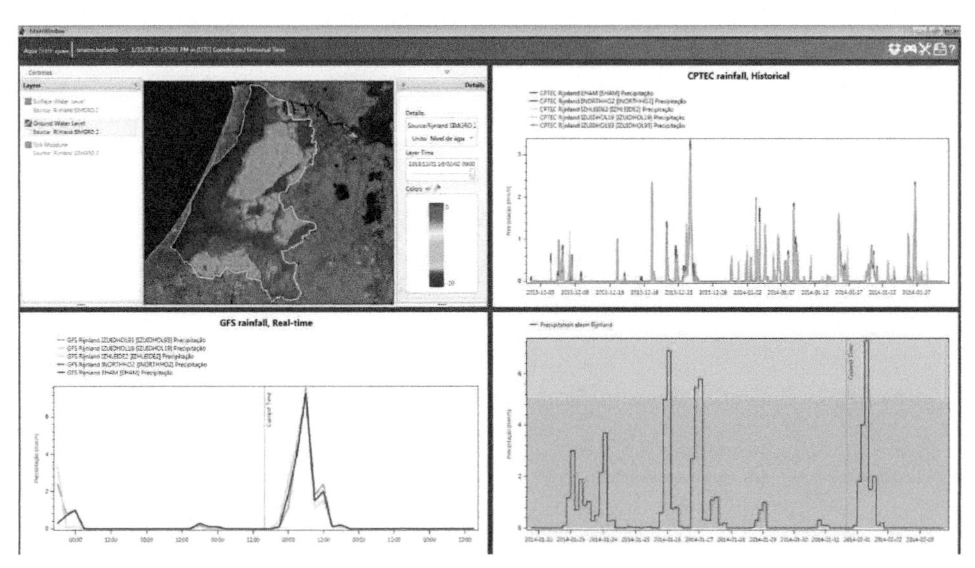

Figure 8-1. Example of a MyWater platform workspace for Rijnland, showing information from a ground water level map, two rainfall data sources, and alert levels

Chapter 9. Conclusions and recommendations

This chapter summarizes the findings and limitations of the study, and provides recommendations for integrating multiple sources of information in hydrological modelling.

9.1 Summary

In this thesis, a framework for effectively integrating the abundance of today's available data sources in hydrological modelling was developed, based on the multi-model ensemble approach. In this framework, multiple available data sources for the same catchment characteristics or hydrometeorological variable, are used to derive catchment parameter values or input time series for the hydrological model. Each unique combination of catchment and input data sources, and data-model integration methods (e.g. merging, data assimilation), thus leads to a different hydrological model simulation result: a new ensemble member. Together, the members form an ensemble of hydrological simulations. By following this approach, all available data sources are used effectively and their information is preserved. The resulting ensemble simulation quantifies part of the data and parameter uncertainty. Assigning weights to the ensemble members determines the probability distribution of the simulation results, and derivatives such as the ensemble mean.

The framework was tested on distributed hydrological modelling of the Rijnland water system in the Netherlands. The hydrological modelling system SIMGRO was used, which is especially suitable for simulating low-land water systems with control structures (weirs, pumping stations), such as Rijnland.

Multiple data-model integration methods were tested: standard direct use as model input, meteorological data merging before using as input, input-data updating based on model outputs (feedback loop), and data assimilation using the particle filter method with residual resampling.

The available data sources for Rijnland comprised three land-use and two soil maps, and two precipitation (in-situ and radar) and two evapotranspiration data sets (in-situ and EO). Following the data-model integration framework, these were used to form 24 unique combinations of catchment parameter values and input time series for the hydrological model, leading to an ensemble simulation of 24 members. As a reference for performance analysis, the discharge records from four main pumping stations were chosen. Various methods to assign weights to the ensemble members were tested to further improve the ensemble simulation performance.

9.2 Conclusions

After analysing and interpreting the Rijnland case study results, the following conclusions can be drawn regarding the specific objectives:

1. *To develop a methodological framework on the basis of the ensemble prediction approach, for incorporation of multiple data sources and multiple integration methods, such as data merging and data assimilation, in hydrological modelling*

   A comprehensive data-model integration framework was developed. In this framework, multiple data sources are used for parameterising catchment characteristics, and multiple data sources are used as hydrometeorological model input, using various data-model integration methods. Each unique combination of catchment characteristics, input data, and data-model integration method, results in another member of an ensemble simulation. Weighting can be applied to further improve the ensemble simulation result and determine the probability distribution and derivatives, such as the ensemble mean.

2. *To develop and validate a distributed hydrological model for a case study, and to analyse its uncertainty*

   A SIMGRO distributed hydrological model of Rijnland has been successfully developed and validated. The model was developed using the data sources usually applied on the case study; the base model. The model was found to produce adequate results with respect to total observed discharge from the Rijnland water system. It should be mentioned however, that due to the presence of a considerable (unrecorded) human influence, not all discharge variability could be reproduced by the model. Although the main validation was against discharge data, secondary validations against polder discharge, ETa, ground water level and surface water level, have been conducted. The validation of the modelled ETa against EO ETa data showed satisfactory results based on agreements in the patterns and reasonable ETa values. On the other hand, the validation of modelled soil moisture against the EO data presented highly deviating results. Both model and satellite data could be wrong, because the hydrological model was not built to simulate the soil moisture accurately and the EO data estimation has a high uncertainty. The high uncertainty in the EO soil moisture data can be seen in the low pattern correlation to the rainfall events. The secondary validation of ground water levels showed acceptable results. Monte Carlo analysis of parametric uncertainty has been carried out and it has been found that the variance of the corresponding model output is low.

3. *To explore the possibility of using model output to fill-in spatial and temporal gaps in EO data*

   An experiment with filling-in EO ETa data with the hydrological model results has been conducted. Feeding back the filled EO ETa maps as input to the hydrological model did not lead to much improvement in the model performance. The main reason can be that there were too many missing data in the EO ETa maps at various locations, and especially in summer periods (when ETa has high impact). As these gaps are filled by the model results, the effect of EO ETa data may mostly be cancelled. The effect of improvement might have been higher if daily EO data had been available (as opposed to the eight-day averaged data used in this study). However, this conclusion is not general but specific to the case study and the simulation period considered and the EO ETa product used.

4. *To implement and test an ensemble EO data assimilation scheme*

   The particle filter method with residual resampling was employed to assimilate EO ETa data into the Rijnland hydrological model. The discharge simulation results showed improvement (10% reduction in PBIAS). The particle method has performed well even with small particle size of 100. The composite map causes a reduction of data, yet the remaining EO ETa information still has led to the improvement of simulated discharge.

5. *To test the data-model integration framework developed, with the case study model and multiple data sources for catchment characteristics and hydrometeorological inputs*

   With the SIMGRO model, ensemble members were constructed with three available land-use maps: LGN6 from the Rijnland Water Board, LANDSAT 5 and GlobCover from satellite products, two soil maps: Dutch database and European database, two observed precipitation data sources: rain gauges and rain radar, and two observed evapotranspiration data sources: weather station and satellite product from Terra/MODIS. Following the data-model integration framework, these data sources were used effectively in 24 different combinations to parameterise and drive the SIMGRO model. The result was an ensemble simulation of discharge from the Rijnland water system from October 2012 to October 2013.

6. *To analyse the performance of the deterministic simulations resulting from the data-model integration framework (individual ensemble members and ensemble mean)*

   Discharge simulations from the individual ensemble members and the ensemble-mean were compared against the measured discharge. The simulations were analysed qualitatively, through hydrographs, and quantitatively with Nash-Sutcliffe efficiency (NSE), percent bias (PBIAS), RSR, and correlation coefficient.

   Visual inspection of the hydrographs showed that the performance of individual members varied in time; hence, no individual member could be identified as uniformly outperforming the others. A sensitivity analysis demonstrated that the parameter uncertainty resulting from different data sources for catchment characteristics, has less effect on the discharge simulation performance than the input uncertainty resulting from different precipitation and evaporation data sources.

   The ensemble mean, based on equal weight of each member, with NSE of 0.85 and PBIAS of 13.1%, was more accurate than most individual members in simulating discharge, including the base model (NSE of 0.81 and PBIAS of 22.2%). Although some individual ensemble members performed better for a certain metric over a certain period, overall, using the ensemble mean led to more accurate flow estimates.

7. *To analyse the performance of the ensemble simulation resulting from the data-model integration framework*

   Together, the multiple hydrological simulation results, being ensemble members, can be assessed probabilistically. The model output ensemble can then be used to derive the empirical probability distributions of simulated discharge, which are compared to the observations. The ensemble showed good performance for identifying events above a range of discharge thresholds. High ROC scores (e.g. 0.80 for the 90th discharge percentile) showed that event threshold exceedances can be simulated with high hit rate and low false alarm rate. The ROC results showed an advantage of using the probability distribution of the ensemble simulation for identifying event threshold exceedance, over using individual simulation results.

8. *To test weighting methods to improve the performance of the ensemble simulation*

   It was found that using a static weighting scheme based on past model performances, rather than using equal weight for each of the members, did not show much further improvement of the ensemble mean. However, using dynamic weighting based on model error during several previous days, has shown more notable improvement over simple averaging. We have tested several dynamic weighting schemes, and dynamically assigning zero weight to a certain percentage of the least performing individual models appeared to be the best.

The main objective of developing and testing a data-model integration framework for incorporating multiple data sources in hydrological modelling, has been achieved. The Rijnland case study results show that the developed framework, based on the multi-model ensemble approach, can be applied effectively, improve discharge simulation, and partially account for the parameter and data uncertainty. When combined with the well-known ensemble methods of parameter sampling, including different model types, and forcing with meteorological ensemble forecasts, a next step can be made towards providing reliable hydrological ensemble simulations and predictions to water managers.

9.3 Limitations of this study

The following limitations of the research should be noted:

1. The framework was applied for one case study, which is Rijnland in the Netherlands. EO data for this research was using the data from the MyWater project and other data sources were not tested.
2. Time period of the analyses is from 2010 to 2013, limiting the training and testing periods, and data sets for determining the performance-based weights.
3. The data assimilation is carried out with one technique, the particle filter. The simulation period for assimilation is only for two and a half months during summer.
4. While this framework is suitable to be used in real-time and forecasting mode, the present study has been limited to simulation mode.
5. The ensemble in this study was constructed using multiple available data sources for catchment characteristics and hydrometeorological variables. The ensemble could be further expanded, using the same framework, by including ensemble weather predictions and various model structures (and their combinations).

These limitations form the basis for developing recommendations for future studies.

9.4 Recommendations

The first recommendation for future work, is to test the framework developed on more case studies, with other water systems, additional or other data sources, additional or other data-model integration methods, and for longer analysis periods. This will provide additional demonstration of the effectiveness of the framework, and allow to continue expanding and improving the framework.

Using additional data sources, such as additional satellite products, data from different in-situ measurement devices, or meteorological reanalysis products, might improve the resulting ensemble simulation. Having more data sources will add complexity to implementations of the data-model integration methods. By using more data sources, uncertainty in the input can be better analysed.

Adding different types of hydrological models, such as a lumped models and data driven models, could also be a valuable expansion of the framework developed. With additional model types, uncertainty in the model structure can be analysed. The additional models will lead to larger ensembles, which can cover a larger part of the total simulation uncertainty.

With respect to the case study of Rijnland, a further filtering can be done to quantify or reduce the human influence in the recorded discharge data. Based on the filtered discharge data, an improvement to the hydrological model could be made. However, in the real case, there will always be human influence on the Rijnland discharge. The improved model might find difficulties in simulating the discharge in Rijnland water system, although the deviations might be explainable.

Experiments using different grid resolutions of the hydrological model can find the best grid-cell size to be used. A lower resolution means faster computational time, while the higher resolution might improve the model. However, in the connection with the earth observation data, the downscaling issue should be noted when using different resolution. Downscaling methods can be included in the framework as data-model integration methods.

More data-model integration methods can be tested, such as Kriging for data merging, and, next to particle filters other data assimilation schemes, e.g. ensemble Kalman filter, variational methods, and nudging. All these methods can be tested within the framework presented, to try and further improve the performance of the resulting hydrological ensemble simulation.

As for the weighting, a more complex weighting scheme, e.g. based on machine learning and committee of models can be implemented. Caution is still needed in periods where one of the data sources temporarily provides unusually low quality data. Further developing dynamic weighting schemes could account for this risk.

The Rijnland case study results show that the presented data-model integration framework, provides an effective way of integrating the variety of hydrometeorological data sources available today. It can be foreseen that combining multiple credible data sources for the same catchment characteristics and input variables, would become part of hydrological modelling practice.

References

Abbott MB, Bathurst JC, Cunge JA, et al (1986) An introduction to the European Hydrological System — Systeme Hydrologique Europeen, "SHE", 2: Structure of a physically-based, distributed modelling system. J Hydrol 87:61–77

AHN (2011) Actueel Hoogtebestand Nederland. http://www.ahn.nl/index.html. Accessed 14 Feb 2014

Ajami N, Duan Q, Sorooshian S (2007) An integrated hydrologic Bayesian multimodel combination framework: Confronting input, parameter, and model structural uncertainty in hydrologic prediction. Water Resour Res

Ajami NK, Duan Q, Gao X, Sorooshian S (2006) Multimodel Combination Techniques for Analysis of Hydrological Simulations: Application to Distributed Model Intercomparison Project Results. J Hydrometeorol 7:755–768 . doi: 10.1175/JHM519.1

Alemu E, Palmer R, Polebitski A, Meaker B (2010) Decision Support System for Optimizing Reservoir Operations Using Ensemble Streamflow Predictions. J Water Resour Plan Manag 137:72–82 . doi: 10.1061/(ASCE)WR.1943-5452.0000088

Alexandridis T, Cherif I, Bilas G, et al (2016) Spatial and Temporal Distribution of Soil Moisture at the Catchment Scale Using Remotely-Sensed Energy Fluxes. Water 8:32 . doi: 10.3390/w8010032

Alexandridis TK, Cherif I, Chemin Y, et al (2009) Integrated Methodology for Estimating Water Use in Mediterranean Agricultural Areas. Remote Sens 1:445–465 . doi: 10.3390/rs1030445

Allen RG, Pereira LS, Raes D, et al (1998) Crop evapotranspiration-Guidelines for computing crop water requirements-FAO Irrigation and drainage paper 56. FAO, Rome 300:D05109

Anaconda Inc. (2018) Miniconda Python package. https://conda.io/miniconda.html. Accessed 4 Sep 2018

Andersen J, Dybkjaer G, Jensen KH, et al (2002) Use of remotely sensed precipitation and leaf area index in a distributed hydrological model. J Hydrol 264:34–50 . doi: http://dx.doi.org/10.1016/S0022-1694(02)00046-X

Andersen J, Refsgaard JC, Jensen KH (2001) Distributed hydrological modelling of the Senegal River Basin — model construction and validation. J Hydrol 247:200–214

Andreadis KM, Lettenmaier DP (2006) Assimilating remotely sensed snow observations into a macroscale hydrology model. Adv Water Resour 29:872–886

Araújo A, Nunes A (2012) MyWater project: GMES specific services for operational water management (LULC, LAI, Evapotranspiration and Soil Moisture). In: Lets Embrace Space Volume II. Publications Office of the European Union, Luxembourg, p 94

Arino O, Bicheron P, Achard F, et al (2008) GlobCover: The most detailed portrait of Earth. Eur Sp Agency Bull 2008:24–31

Atkinson PM (2012) Downscaling in remote sensing. Int J Appl Earth Obs Geoinf

Aubert D, Loumagne C, Oudin L (2003) Sequential assimilation of soil moisture and streamflow data in a conceptual rainfall–runoff model. J Hydrol 280:145–161

Bai P, Liu X, Liu C (2018) Improving hydrological simulations by incorporating GRACE data for model calibration. J Hydrol 557:291–304 . doi: 10.1016/J.JHYDROL.2017.12.025

Berne A, Krajewski WF (2012) Radar for hydrology: Unfulfilled promise or unrecognized potential? Adv Water Resour

Beven K (2001) How far can we go in distributed hydrological modelling? Hydrol Earth Syst Sci 5:1–12

Beven K (2006) A manifesto for the equifinality thesis. J Hydrol 320:18–36

Beven K (1989) Changing ideas in hydrology — The case of physically-based models. J Hydrol 105:157–172

Beven K, Binley A (2014) GLUE: 20 years on. Hydrol Process 28:5897–5918 . doi: 10.1002/hyp.10082

Beven K, Freer J (2001) Equifinality, data assimilation, and uncertainty estimation in mechanistic modelling of complex environmental systems using the GLUE methodology. J Hydrol 249:11–29

Blyth K (1993) The use of microwave remote sensing to improve spatial parameterization of hydrological models. J Hydrol 152:103–129

Boegh E, Thorsen M, Butts MB, et al (2004) Incorporating remote sensing data in physically based distributed agro-hydrological modelling. J Hydrol 287:279–299

Buizza R, Milleer M, Palmer TN (1999) Stochastic representation of model uncertainties in the ECMWF ensemble prediction system. Q J R Meteorol Soc 125:2887–2908 . doi: 10.1002/qj.49712556006

Butts MB, Payne JT, Kristensen M, Madsen H (2004) An evaluation of the impact of model structure on hydrological modelling uncertainty for streamflow simulation. J Hydrol 298:242–266

Cane D, Milelli M (2010) Multimodel SuperEnsemble technique for quantitative precipitation forecasts in Piemonte region. Nat Hazards Earth Syst Sci 10:265–273 . doi: 10.5194/nhess-10-265-2010

Canonical Ltd (2018) Ubuntu Desktop. https://www.ubuntu.com/. Accessed 4 Apr 2018

Chen JM, Chen X, Ju W, Geng X (2005) Distributed hydrological model for mapping evapotranspiration using remote sensing inputs. J Hydrol 305:15–39

Cheng L, AghaKouchak A (2015) A methodology for deriving ensemble response from multimodel simulations. J Hydrol 522:49–57 . doi: 10.1016/J.JHYDROL.2014.12.025

Cherif I, Alexandridis TK, Jauch E, et al (2015) Improving remotely sensed actual evapotranspiration estimation with raster meteorological data. Int J Remote Sens 36:4606–4620 . doi: 10.1080/01431161.2015.1084439

Chiang Y-M, Hsu K-L, Chang F-J, et al (2007) Merging multiple precipitation sources for flash flood forecasting. J Hydrol 340:183–196

Clark MP, Bierkens MFP, Samaniego L, et al (2017) The evolution of process-based hydrologic models: historical challenges and the collective quest for physical realism. Hydrol Earth Syst Sci 21:. doi: 10.5194/hess-21-3427-2017

Clark MP, Slater AG, Barrett AP, et al (2006) Assimilation of snow covered area information into hydrologic and land-surface models. Adv Water Resour 29:1209–1221 . doi: DOI 10.1016/j.advwatres.2005.10.001

Cole SJ, Moore RJ (2008) Hydrological modelling using raingauge- and radar-based estimators of areal rainfall. J Hydrol 358:159–181

Cole TM, Wells SA (2000) User Manual-CE-QUAL-W2: A Two Dimensional, Laterally Averaged Hydrodynamic and Water Quality Model, Version 3.0. US Army Corps Eng Washington, DC

CPTEC/INPE (2014) Center for Weather Forecasting and Climate Research (Centro de Previsão de Tempo e Estudos Climáticos) CPTEC/INPE. http://www.cptec.inpe.br/. Accessed 12 Feb 2014

Crochet P (2009) Enhancing radar estimates of precipitation over complex terrain using information derived from an orographic precipitation model. J Hydrol 377:417–433

Crow WT, Bolten JD (2007) Estimating precipitation errors using spaceborne surface soil moisture retrievals. Geophys Res Lett 34:

Crow WT, van den Berg MJ, Huffman GJ, Pellarin T (2011) Correcting rainfall using satellite-based surface soil moisture retrievals: The Soil Moisture Analysis Rainfall Tool (SMART). Water Resour Res 47:

Curnel Y, de Wit AJWW, Duveiller G, Defourny P (2011) Potential performances of remotely sensed LAI assimilation in WOFOST model based on an OSS Experiment. Agric For Meteorol 151:1843–1855 . doi: 10.1016/j.agrformet.2011.08.002

de Bruin HAR, Lablans WN (1998) Reference crop evapotranspiration determined with a modified Makkink equation. Hydrol Process 12:1053–1062 . doi: 10.1002/(SICI)1099-1085(19980615)12:7<1053::AID-HYP639>3.0.CO;2-E

De Gruijter JJ, Horst JBF van der, Heuvelink GBM, et al (2004) Grondwater opnieuw op de kaart. Wageningen, Alterra report 915

De Lannoy GJM, Reichle RH, Arsenault KR, et al (2012) Multiscale assimilation of Advanced Microwave Scanning Radiometer–EOS snow water equivalent and Moderate Resolution Imaging Spectroradiometer snow cover fraction observations in northern Colorado. Water Resour Res 48:W01522 . doi: 10.1029/2011WR010588

de Vries F, de Groot WJM, Hoogland T, Denneboorm J (2003) De Bodemkaart van Netherland digitaal;. Toelichting bij inhound, actualiteit en methodiek en korte beschrijving van additionel informatie. Wageningen

Dechant C, Moradkhani H (2011) Radiance data assimilation for operational snow and streamflow forecasting. Adv Water Resour 34:351–364 . doi: 10.1016/j.advwatres.2010.12.009

Devia GK, Ganasri BP, Dwarakish GS (2015) A Review on Hydrological Models. Aquat Procedia 4:1001–1007 . doi: 10.1016/J.AQPRO.2015.02.126

Diepen CA, Wolf J, Keulen H, Rappoldt C (1989) WOFOST: a simulation model of crop production. Soil Use Manag 5:16–24 . doi: 10.1111/j.1475-2743.1989.tb00755.x

Dino loket (2013) Ondergrondgegevens. https://www.dinoloket.nl/ondergrondgegevens. Accessed 3 May 2013

Dogulu N, López López P, Solomatine DP, et al (2015) Estimation of predictive hydrologic uncertainty using the quantile regression and UNEEC methods and their comparison on contrasting catchments. Hydrol Earth Syst Sci 19:3181–3201 . doi: 10.5194/hess-19-3181-2015

Duan Q, Ajami NK, Gao X, Sorooshian S (2007) Multi-model ensemble hydrologic prediction using Bayesian model averaging. Adv Water Resour 30:1371–1386 . doi: 10.1016/j.advwatres.2006.11.014

Ebert E (2001) Ability of a poor man's ensemble to predict the probability and distribution of precipitation. Mon Weather Rev

EC (2003) European Soil Database (distribution version v2.0)

ECMWF (2015) ECMWF | Advancing global NWP through co-operation. http://www.ecmwf.int/. Accessed 18 Dec 2015

Einfalt T, Arnbjerg-Nielsen K, Golz C, et al (2004) Towards a roadmap for use of radar rainfall data in urban drainage. J Hydrol 299:186–202

Evensen G (1994) Sequential data assimilation with a nonlinear quasi-geostrophic model using Monte Carlo methods to forecast error statistics. J Geophys Res Ser 99:10

Evensen G (2003) The ensemble Kalman filter: Theoretical formulation and practical implementation. Ocean Dyn 53:343–367

Evora ND, Coulibaly P (2009) Recent advances in data-driven modeling of remote sensing applications in hydrology. J Hydroinformatics 11:194–201 . doi: DOI 10.2166/hydro.2009.036

Feddes RA, Kowalik PJ, Zaradny H (1978) Simulation of field water use and crop yields. Simulation monographs. Wageningen

Fox D, Burgard W, Dellaert F, Thrun S (1999) Monte carlo localization: Efficient position estimation for mobile robots. AAAI/IAAI 1999:343–349

Georgakakos KP, Seo D-J, Gupta H, et al (2004) Towards the characterization of streamflow simulation uncertainty through multimodel ensembles. J Hydrol 298:222–241

Golembesky K, Sankarasubramanian A, Devineni N (2009) Improved Drought Management of Falls Lake Reservoir: Role of Multimodel Streamflow Forecasts in Setting up Restrictions. J Water Resour Plan Manag 135:188–197 . doi: 10.1061/(ASCE)0733-9496(2009)135:3(188)

Gordon NJ, Salmond DJ, Smith AFM (1993) Novel approach to nonlinear/non-Gaussian Bayesian state estimation. IEE Proc F (Radar Signal Process 140:107–113(6)

Goudenhoofdt E, Delobbe L (2009) Evaluation of radar-gauge merging methods for quantitative precipitation estimates. Hydrol Earth Syst Sci 13:195–203 . doi: 10.5194/hess-13-195-2009

Gourley JJ, Vieux BE (2006) A method for identifying sources of model uncertainty in rainfall-runoff simulations. J Hydrol 327:68–80

Haberlandt U (2007) Geostatistical interpolation of hourly precipitation from rain gauges and radar for a large-scale extreme rainfall event. J Hydrol 332:144–157

Han E, Merwade V, Heathman GC (2012) Implementation of surface soil moisture data assimilation with watershed scale distributed hydrological model. J Hydrol 416–417:98–117

Hartanto IM, Almeida C, Alexandridis TK, et al (2015) Merging earth observation data, weather predictions, in-situ measurements and hydrological models for water information services. Environ Eng Manag J 14:2031–2042

Hazeu GW, Schuiling C, Dorland GJ, Gijsbertse HA (2010) Landelijk Grondgebruiksbestand Nederland versie 6 (LGN6). Wageningen

Hiemstra PH, Karssenberg D, van Dijk A (2011) Assimilation of observations of radiation level into an atmospheric transport model: A case study with the particle filter and the ETEX tracer dataset. Atmos Environ 45:6149–6157 . doi: 10.1016/j.atmosenv.2011.08.024

Hiep NH, Luong ND, Viet Nga TT, et al (2018) Hydrological model using ground- and satellite-based data for river flow simulation towards supporting water resource management in the Red River Basin, Vietnam. J Environ Manage 217:346–355 . doi: 10.1016/J.JENVMAN.2018.03.100

Huber WC, Rossman LA, Dickinson RA (2005) EPA storm water management model SWMM 5. 0. Watershed Model CRC Press Boca Raton, FL 339–361

Hue C, Le Cadre J-P, Perez P (2002) Tracking multiple objects with particle filtering. Aerosp Electron Syst IEEE Trans 38:791–812 . doi: 10.1109/TAES.2002.1039400

Huffman GJ (1995) Global precipitation estimates based on a technique for combining satellite-based estimates, rain gauge analysis, and NWP model precipitation information

Hydronet (2014) Hydronet. http://www.hydronet.nl/

Immerzeel WW, Droogers P (2008) Calibration of a distributed hydrological model based on satellite evapotranspiration. J Hydrol 349:411–424 . doi: DOI 10.1016/j.jhydrol.2007.11.017

Irmak A, Kamble B (2009) Evapotranspiration data assimilation with genetic algorithms and SWAP model for on-demand irrigation. Irrig Sci 28:101–112 . doi: 10.1007/s00271-009-0193-9

Jackson TJ, Schmugge J, Engman ET (1996) Remote sensing applications to hydrology: Soil moisture. Hydrol Sci Journal-Journal Des Sci Hydrol 41:517–530

Jaun S, Germann U, Walser A, Fundel F (2011) Superposition of three sources of uncertainties in operational flood forecasting chains. Atmos Res 100:246–262 . doi: 10.1016/J.ATMOSRES.2010.12.005

Jian J, Ryu D, Costelloe JF, Su C-H (2017) Towards hydrological model calibration using river level measurements. J Hydrol Reg Stud 10:95–109 . doi: 10.1016/J.EJRH.2016.12.085

Jiang S, Ren L, Hong Y, et al (2012) Comprehensive evaluation of multi-satellite precipitation products with a dense rain gauge network and optimally merging their simulated hydrological flows using the Bayesian model averaging method. J Hydrol 452–453:213–225

Karssenberg D (2010) Land surface process modelling with PCRaster Python

Karssenberg D, Jong K De (2005) Dynamic environmental modelling in GIS: 2. Modelling error propagation. Int J Geogr Inf Sci 19:623–637 . doi: 10.1080/13658810500104799

Karssenberg D, Schmitz O, Salamon P, et al (2010) A software framework for construction of process-based stochastic spatio-temporal models and data assimilation. Environ Model Softw 25:489–502 . doi: 10.1016/j.envsoft.2009.10.004

Kayastha N, Ye J, Fenicia F, et al (2013) Fuzzy committees of specialized rainfall-runoff models: further enhancements and tests. Hydrol Earth Syst Sci 17:4441–4451 . doi: 10.5194/hess-17-4441-2013

KNMI (2014) Royal Netherlands Meteorological Institute. https://data.knmi.nl/portal/KNMI-DataCentre.html

Krajewski WF (1987) Cokriging radar-rainfall and rain gage data. J Geophys Res 92:9571 . doi: 10.1029/JD092iD08p09571

Krajewski WF, Anderson MC, Eichinger WE, et al (2006) A remote sensing observatory for hydrologic sciences: A genesis for scaling to continental hydrology. Water Resour Res 42:

Kumari M, Singh CK, Basistha A (2017) Clustering Data and Incorporating Topographical Variables for Improving Spatial Interpolation of Rainfall in Mountainous Region. Water Resour Manag 31:425–442 . doi: 10.1007/s11269-016-1534-0

Kurtz W, Lapin A, Schilling OS, et al (2017) Integrating hydrological modelling, data assimilation and cloud computing for real-time management of water resources. Environ Model Softw 93:418–435 . doi: 10.1016/J.ENVSOFT.2017.03.011

Lanza LG, Schultz GA, Barrett EC (1997) Remote sensing in hydrology: Some downscaling and uncertainty issues. Phys Chem Earth 22:215–219

Lee H, Seo D-J, Koren V (2011) Assimilation of streamflow and in situ soil moisture data into operational distributed hydrologic models: Effects of uncertainties in the data and initial model soil moisture states. Adv Water Resour 34:1597–1615 . doi: 10.1016/j.advwatres.2011.08.012

Li H, Zhang Y, Chiew FHS, Xu S (2009) Predicting runoff in ungauged catchments by using Xinanjiang model with MODIS leaf area index. J Hydrol 370:155–162

Li M, Shao Q (2010) An improved statistical approach to merge satellite rainfall estimates and raingauge data. J Hydrol 385:51–64

Li Y, Grimaldi S, Pauwels VRN, Walker JP (2018) Hydrologic model calibration using remotely sensed soil moisture and discharge measurements: The impact on predictions at gauged and ungauged locations. J Hydrol 557:897–909 . doi: 10.1016/J.JHYDROL.2018.01.013

Liao K, Xua F, Zheng J, et al (2014) Using different multimodel ensemble approaches to simulate soil moisture in a forest site with six traditional pedotransfer functions. Environ Model Softw 57:27–32 . doi: 10.1016/j.envsoft.2014.03.016

Lievens H, Tomer SK, Al Bitar A, et al (2015) SMOS soil moisture assimilation for improved hydrologic simulation in the Murray Darling Basin, Australia. Remote Sens Environ 168:146–162 . doi: 10.1016/j.rse.2015.06.025

Liu JS, Chen R (1998) Sequential Monte Carlo Methods for Dynamic Systems. J Am Stat Assoc 93:1032–1044 . doi: 10.1080/01621459.1998.10473765

Liu S, Shao Y, Yang C, et al (2012a) Improved regional hydrologic modelling by assimilation of streamflow data into a regional hydrologic model. Environ Model & Softw 31:141–149 . doi: 10.1016/j.envsoft.2011.12.005

Liu Y, Gupta H (2007) Uncertainty in hydrologic modeling: Toward an integrated data assimilation framework. Water Resour Res

Liu Y, Weerts a. H, Clark M, et al (2012b) Advancing data assimilation in operational hydrologic forecasting: progresses, challenges, and emerging opportunities. Hydrol Earth Syst Sci 16:3863–3887 . doi: 10.5194/hess-16-3863-2012

Loaiza Usuga JC, Pauwels VRNN (2008) Calibration and multiple data set-based validation of a land surface model in a mountainous Mediterranean study area. J Hydrol 356:223–233 . doi: 10.1016/j.jhydrol.2008.04.018

Mantovan P, Todini E (2006) Hydrological forecasting uncertainty assessment: Incoherence of the GLUE methodology. J Hydrol 330:368–381

Margulis SA, McLaughlin D, Entekhabi D, Dunne S (2002) Land data assimilation and estimation of soil moisture using measurements from the Southern Great Plains 1997 Field Experiment. Water Resour Res 38:35-1-35–18 . doi: 10.1029/2001WR001114

Martens B, Cabus P, De Jongh I, Verhoest NEC (2013) Merging weather radar observations with ground-based measurements of rainfall using an adaptive multiquadric surface fitting algorithm. J Hydrol 500:84–96 . doi: 10.1016/j.jhydrol.2013.07.011

Matgen P, Montanari M, Hostache R, et al (2010) Towards the sequential assimilation of SAR-derived water stages into hydraulic models using the Particle Filter: proof of concept. Hydrol Earth Syst Sci 14:1773–1785 . doi: 10.5194/hess-14-1773-2010

Mazzoleni M, Alfonso L, Chacon-Hurtado J, Solomatine D (2015) Assimilating uncertain, dynamic and intermittent streamflow observations in hydrological models. Adv Water Resour 83:323–339 . doi: 10.1016/J.ADVWATRES.2015.07.004

Mesinger F, Chou SC, Gomes JL, et al (2012) An upgraded version of the ETA model. Meteorol Atmos Phys 116:63–79

Michaelides S, Levizzani V, Anagnostou E, et al (2009) Precipitation: Measurement, remote sensing, climatology and modeling. Atmos Res 94:512–533

Moradkhani H (2008) Hydrologic remote sensing and land surface data assimilation. Sensors 8:2986–3004 . doi: Doi 10.3390/S8052986

Moradkhani H, Hsu K-L, Gupta H, Sorooshian S (2005) Uncertainty assessment of hydrologic model states and parameters: Sequential data assimilation using the particle filter. Water Resour Res 41:n/a-n/a . doi: 10.1029/2004wr003604

Moriasi D, Arnold J (2007) Model evaluation guidelines for systematic quantification of accuracy in watershed simulations. Trans ASABE 503 885–900

MyWater project (2014) MyWater EU FP7 Project. http://mywater-fp7.eu/. Accessed 5 Aug 2014

Nagler T, Rott H, Malcher P, Müller F (2008) Assimilation of meteorological and remote sensing data for snowmelt runoff forecasting. Remote Sens Environ 112:1408–1420

Neitsch S, Arnold J, Kiniry J, et al (2005) SWAT theoretical documentation version 2005. Blackl Res Center, ...

Nerini D, Zulkafli Z, Wang L-P, et al (2015) A Comparative Analysis of TRMM–Rain Gauge Data Merging Techniques at the Daily Time Scale for Distributed Rainfall–Runoff Modeling Applications. J Hydrometeorol 16:2153–2168 . doi: 10.1175/JHM-D-14-0197.1

Newman AJ, Clark MP, Craig J, et al (2015) Gridded Ensemble Precipitation and Temperature Estimates for the Contiguous United States. J Hydrometeorol 16:2481–2500 . doi: 10.1175/JHM-D-15-0026.1

NHI (2014) Nationaal Hydrologisch Instrumentarium. http://www.nhi.nu/. Accessed 5 Aug 2014

Noh SJ, Tachikawa Y, Shiiba M, Kim S (2011) Applying sequential Monte Carlo methods into a distributed hydrologic model: lagged particle filtering approach with regularization. Hydrol Earth Syst Sci Discuss 8:3383–3420 . doi: 10.5194/hessd-8-3383-2011

Nunes A, Araújo A, Alexandridis T, et al (2013) Effects of different scale land cover maps in watershed modelling. In: EGU General Assembly Conference Abstracts. p 11990

Olioso a., Inoue Y, Ortega-FARIAS S, et al (2005) Future directions for advanced evapotranspiration modeling: Assimilation of remote sensing data into crop simulation models and SVAT models. Irrig Drain Syst 19:377–412 . doi: 10.1007/s10795-005-8143-z

Pappenberger F, Ramos MH, Cloke HL, et al (2015) How do I know if my forecasts are better? Using benchmarks in hydrological ensemble prediction. J Hydrol 522:697–713 . doi: 10.1016/J.JHYDROL.2015.01.024

Pauwels VRN, Hoeben R, Verhoest NEC, De Troch FP (2001) The importance of the spatial patterns of remotely sensed soil moisture in the improvement of discharge predictions for small-scale basins through data assimilation. J Hydrol 251:88–102

Pauwels VRN, Verhoest NEC, De Lannoy GJM, et al (2007) Optimization of a coupled hydrology-crop growth model through the assimilation of observed soil moisture and leaf area index values using an ensemble Kalman filter. Water Resour Res 43:n/a-n/a . doi: 10.1029/2006WR004942

Perrin C, Anctil F, Seiller G (2012) Multimodel evaluation of twenty lumped hydrological models under contrasted climate conditions. Hydrol Earth Syst Sci 16:1171–1189 . doi: 10.5194/hess-1116-1171-2012

Price R, van der Wielen J, Velickov S, Galvao D (2014) A 2D simulation model for urban flood management. In: EGU General Assembly Conference Abstracts. p 12440

PuTTY (2017) PuTTY. https://www.putty.org/. Accessed 12 Dec 2017

Qin C, Jia Y, Su Z (Bob., et al (2008) Integrating Remote Sensing Information Into A Distributed Hydrological Model for Improving Water Budget Predictions in Large-scale Basins through Data Assimilation. Sensors 8:4441–4465 . doi: 10.3390/s8074441

Regonda S, Zagona E, Rajagopalan B (2011) Prototype Decision Support System for Operations on the Gunnison Basin with Improved Forecasts. J Water Resour Plan Manag 137:428–438 . doi: 10.1061/(ASCE)WR.1943-5452.0000133

Reichle RH (2008) Data assimilation methods in the Earth sciences. Adv Water Resour 31:1411–1418

Reichle RH, McLaughlin DB, Entekhabi D (2002) Hydrologic data assimilation with the ensemble Kalman filter. Mon Weather Rev 130:103–114 . doi: 10.1175/1520-0493(2002)130<0103:HDAWTE>2.0.CO;2

Rientjes THM, Muthuwatta LP, Bos MG, et al (2013) Multi-variable calibration of a semi-distributed hydrological model using streamflow data and satellite-based evapotranspiration. J Hydrol 505:276–290 . doi: 10.1016/j.jhydrol.2013.10.006

Rijnland Water Board (2014) Rijnland Water Board. http://www.rijnland.net/. Accessed 5 Aug 2014

Sahoo GB, Ray C, De Carlo EH (2006) Calibration and validation of a physically distributed hydrological model, MIKE SHE, to predict streamflow at high frequency in a flashy mountainous Hawaii stream. J Hydrol 327:94–109

Salvadore E, Bronders J, Batelaan O (2015) Hydrological modelling of urbanized catchments: A review and future directions. J Hydrol 529:62–81 . doi: 10.1016/J.JHYDROL.2015.06.028

Schaake JC, Hamill TM, Buizza R, Clark M (2007) HEPEX: The Hydrological Ensemble Prediction Experiment. Bull Am Meteorol Soc 88:1541–1548 . doi: 10.1175/BAMS-88-10-1541

Schmugge TJ, Kustas WP, Ritchie JC, et al (2002) Remote sensing in hydrology. Adv Water Resour 25:1367–1385

Schultz GA, Engman ET (2000) Remote sensing in hydrology and water management : with 184 figures and 22 tables. Springer, Berlin [etc.]

Schuurmans JM, Troch PA, Veldhuizen AA, et al (2003) Assimilation of remotely sensed latent heat flux in a distributed hydrological model. Adv Water Resour 26:151–159 . doi: 10.1016/S0309-1708(02)00089-1

Scott CA, Bastiaanssen WGM, Ahmad M-D (2003) Mapping root zone soil moisture using remotely sensed optical imagery. J Irrig Drain Eng 129:326–335 . doi: 10.1061/(ASCE)0733-9437(2003)129:5(326)

Seo D-J, Breidenbach J., Johnson E. (1999) Real-time estimation of mean field bias in radar rainfall data. J Hydrol 223:131–147 . doi: 10.1016/S0022-1694(99)00106-7

Seo Y, Kim S, Singh VP (2015) Estimating Spatial Precipitation Using Regression Kriging and Artificial Neural Network Residual Kriging (RKNNRK) Hybrid Approach. Water Resour Manag 29:2189–2204 . doi: 10.1007/s11269-015-0935-9

Shamseldin AY, O'Connor KM, Liang GC (1997) Methods for combining the outputs of different rainfall–runoff models. J Hydrol 197:203–229 . doi: http://dx.doi.org/10.1016/S0022-1694(96)03259-3

Shrestha DL, Kayastha N, Solomatine D, Price R (2014) Encapsulation of parametric uncertainty statistics by various predictive machine learning models: MLUE method. J Hydroinformatics 16:95

Shrestha DL, Solomatine DP (2008) Data-driven approaches for estimating uncertainty in rainfall-runoff modelling. Int J River Basin Manag 6:109–122

Skinner CJ, Bellerby TJ, Greatrex H, Grimes DIF (2015) Hydrological modelling using ensemble satellite rainfall estimates in a sparsely gauged river basin: The need for whole-ensemble calibration. J Hydrol 522:110–122 . doi: 10.1016/J.JHYDROL.2014.12.052

Sokol Z (2003) Utilization of regression models for rainfall estimates using radar-derived rainfall data and rain gauge data. J Hydrol 278:144–152 . doi: 10.1016/S0022-1694(03)00139-2

Sokol Z (2011) Assimilation of extrapolated radar reflectivity into a NWP model and its impact on a precipitation forecast at high resolution. Atmos Res 100:201–212 . doi: 10.1016/j.atmosres.2010.09.008

Solomatine DP (2006) Data-Driven Modeling and Computational Intelligence Methods in Hydrology. In: Encyclopedia of Hydrological Sciences. John Wiley & Sons, Ltd

Stisen S, Jensen KH, Sandholt I, Grimes DIF (2008) A remote sensing driven distributed hydrological model of the Senegal River basin. J Hydrol 354:131–148

Strauch M, Bernhofer C, Koide S, et al (2012) Using precipitation data ensemble for uncertainty analysis in SWAT streamflow simulation. J Hydrol 414–415:413–424

Sun X, Mein RG, Keenan TD, Elliott JF (2000) Flood estimation using radar and raingauge data. J Hydrol 239:4–18

Surfsara (2014) HPC Cloud Surfsara. https://userinfo.surfsara.nl/systems/hpc-cloud. Accessed 4 Aug 2014

Tapiador FJ, Turk FJ, Petersen W, et al (2012) Global precipitation measurement: Methods, datasets and applications. Atmos Res 104–105:70–97

Teegavarapu RS V, Chandramouli V (2005) Improved weighting methods, deterministic and stochastic data-driven models for estimation of missing precipitation records. J Hydrol 312:191–206

Thiboult A, Anctil F, Boucher M-A (2016) Accounting for three sources of uncertainty in ensemble hydrological forecasting. Hydrol Earth Syst Sci 20:1809–1825 . doi: 10.5194/hess-20-1809-2016

Todini E (2007) Hydrological catchment modelling: past, present and future. Hydrol Earth Syst Sci 11:468–482

Tóth B, Weynants M, Nemes A, et al (2015) New generation of hydraulic pedotransfer functions for Europe. Eur J Soil Sci 66:226–238 . doi: 10.1111/ejss.12192

Trancoso AR, Braunschweig F, Chambel Leitão P, et al (2009) An advanced modelling tool for simulating complex river systems. Sci Total Environ 407:3004–3016 . doi: 10.1016/j.scitotenv.2009.01.015

van Andel S, Price R, Lobbrecht A, et al (2014) Framework for Anticipatory Water Management: Testing for Flood Control in the Rijnland Storage Basin. J Water Resour Plan Manag 140:533–542 . doi: 10.1061/(ASCE)WR.1943-5452.0000254

van Andel SJ, Price R, Lobbrecht A, van Kruiningen F (2010) Modeling Controlled Water Systems. J Irrig Drain Eng 136:392–404 . doi: Doi 10.1061/(Asce)Ir.1943-4774.0000159

van Andel SJ, Price RK, Lobbrecht AH, et al (2008) Ensemble Precipitation and Water-Level Forecasts for Anticipatory Water-System Control. J Hydrometeorol 9:776–788 . doi: 10.1175/2008JHM971.1

Van Coillie FMB, Lievens H, Joos I, et al (2011) Training neural networks on artificially generated data: a novel approach to SAR speckle removal. Int J Remote Sens 32:3405–3425 . doi: 10.1080/01431161003749436 van Leeuwen PJ (2003) A Variance Minimizing Filter for Large Scale Applications. Mon Weather Rev 131:2071–2084 . doi: 10.1175/1520-0493(2003)131<2071:AVFFLA>2.0.CO;2

Van Walsum PE V (2011) SIMGRO 7.2.0 ; User's guide. Alterra-Wageningen UR, Wageningen

van Walsum PE V, Veldhuizen AA (2011) Integration of models using shared state variables: Implementation in the regional hydrologic modelling system SIMGRO. J Hydrol 409:363–370

Van Walsum PE V, Veldhuizen AA, Groenendijk P (2011) SIMGRO 7.2.0, Theory and model implementation. Alterra-Wageningen UR, Wageningen

Vasiloff S V, Howard KW, Zhang J (2009) Difficulties with Correcting Radar Rainfall Estimates Based on Rain Gauge Data: A Case Study of Severe Weather in Montana on 16–17 June 2007. Weather Forecast 24:1334–1344 . doi: 10.1175/2009WAF2222154.1

Vazifedoust M, van Dam JC, Bastiaanssen WGM, Feddes RA (2009) Assimilation of satellite data into agrohydrological models to improve crop yield forecasts. Int J Remote Sens 30:2523–2545 . doi: 10.1080/01431160802552769

Velasco-Forero CA, Sempere-Torres D, Cassiraga EF, Jaime Gómez-Hernández J (2009) A non-parametric automatic blending methodology to estimate rainfall fields from rain gauge and radar data. Adv Water Resour 32:986–1002 . doi: 10.1016/j.advwatres.2008.10.004

Velázquez JA, Anctil F, Perrin C (2010) Performance and reliability of multimodel hydrological ensemble simulations based on seventeen lumped models and a thousand catchments. Hydrol Earth Syst Sci 14:2303–2317 . doi: 10.5194/hess-14-2303-2010

Verkade JS, Brown JD, Reggiani P, Weerts AH (2013) Post-processing ECMWF precipitation and temperature ensemble reforecasts for operational hydrologic forecasting at various spatial scales. J Hydrol 501:73-91. doi: 10.1016/j.jhydrol.2013.07.039

Vrugt JA, Diks CGH, Gupta H V, et al (2005) Improved treatment of uncertainty in hydrologic modeling: Combining the strengths of global optimization and data assimilation. Water Resour Res 41:

Weerts AH, El Serafy GYH (2006) Particle filtering and ensemble Kalman filtering for state updating with hydrological conceptual rainfall-runoff models. Water Resour Res 42:n/a-n/a . doi: 10.1029/2005WR004093

Weynants M, Montanarella L, Tóth G, et al (2013) European HYdropedological Data Inventory (EU-HYDI). EUR – Scientific and Technical Research series – ISSN 1831-9424, Luxembourg

WMO (2012) Guidelines on Ensemble Prediction Systems and Forecasting. World Meteorological Organization (WMO)

Wu CL, Chau KW (2010) Data-driven models for monthly streamflow time series prediction. Eng Appl Artif Intell 23:1350–1367 . doi: 10.1016/j.engappai.2010.04.003

X2Go (2018) X2Go everywhere@home. https://wiki.x2go.org/doku.php. Accessed 4 Sep 2018

Xie X, Zhang D (2010) Data assimilation for distributed hydrological catchment modeling via ensemble Kalman filter. Adv Water Resour 33:678–690 . doi: 10.1016/j.advwatres.2010.03.012

Yan H, Moradkhani H (2016) Combined assimilation of streamflow and satellite soil moisture with the particle filter and geostatistical modeling. Adv Water Resour 94:364–378 . doi: 10.1016/J.ADVWATRES.2016.06.002

Yu J (2005) On leverage in a stochastic volatility model. J Econom 127:165–178 . doi: 10.1016/j.jeconom.2004.08.002

Zalachori I, Ramos M-H, Garon R, Mathevet T, Gailhard J (2012) Statistical processing of forecasts for hydrological ensemble prediction: a comparative study of different bias correction strategies. Advances in Science and Research 8:135-141

Zhou SK, Chellappa R, Moghaddam B (2004) Visual tracking and recognition using appearance-adaptive models in particle filters. Image Process IEEE Trans 13:1491–1506 . doi: 10.1109/TIP.2004.836152

Zou L, Zhan C, Xia J, et al (2017) Implementation of evapotranspiration data assimilation with catchment scale distributed hydrological model via an ensemble Kalman Filter. J Hydrol 549:685–702 . doi: 10.1016/J.JHYDROL.2017.04.036

Samenvatting

De beschikbaarheid van Earth Observation en Numerical Weather Prediction gegevens voor hydrologische modellering en waterbeheer is sterk toegenomen. Met deze toename is het vandaag de dag zo dat voor één en dezelfde variabele, schattingen van twee of meer informatiebronnen beschikbaar kunnen zijn. Neerslaggegevens kunnen bijvoorbeeld afkomstig zijn van grondstations, weerradar, satellieten, en weermodellen. Landgebruik kan afkomstig zijn van veldonderzoek, satellietgegevens, of een combinatie van die twee. Elk van deze gegevensbronnen levert een schatting van een stroomgebiedskarakteristiek en daaraan gerelateerde parameterwaarden voor een hydrologisch model op, of van een hydrometeorologische variabele. Schattingen van de verschillende gegevensbronnen (data-producten) verschillen van elkaar in kwantiteit of in ruimtelijke en temporele variatie. Het is niet altijd mogelijk om te beoordelen welk data-product het meest nauwkeuring is. Een bepaald product kan een slechte schatting geven in de ene situatie en voor andere situaties juist wel nauwkeurig zijn. Toch wordt voor hydrologisch modellering meestal een bepaalde set producten voor stroomgebiedskarakteristieken en invoertijdreeksen geselecteerd, waarbij mogelijk andere relevante gegevensbronnen worden genegeerd. Eén van de redenen kan zijn dat er voor hydrologische modellering, ondanks uitgebreid onderzoek en ontwikkeling naar integratiemethodes voor een deel van de beschikbare gegevensbronnen, geen gedegen raamwerk is voor integratie van gegevens en modellen (data-model integration) dat uit gaat van het bestaan en gebruik van meerdere gegevensbronnen naast elkaar.

Het hoofddoel van deze dissertatie is zo'n raamwerk te ontwikkelen en deze vervolgens te testen in een praktijkonderzoek .

Het ontwikkelde raamwerk is gebaseerd op de methodiek van ensemble voorspellingen (Ensemble Prediction). Ensemble Prediction is een vorm van probabilistisch voorspellen waarbij in plaats van één (deterministische) voorspelling voor een bepaalde tijd en plaats te geven, meerdere voorspellingen (members) worden gegeven. De alternatieve voorspellingen worden gegenereerd om de effecten van onzekerheden in beginwaarden, parameterwaarden, modelinvoer, en modelstructuur weer te geven. Door een kans van optreden toe te kennen aan elk van de voorspellingen, members, kan de kansverdeling van de gehele ensemble voorspelling worden bepaald. Ensemble technieken zijn met name in de meteorologie, voor modelmatige weersvoorspellingen, ontwikkeld en als eerste op grote schaal toegepast. Daarbij werd bijvoorbeeld een ensemble voorspelling gegenereerd door steeds de beginwaarden van een meteorologisch model een klein beetje aan te passen en het model nog een keer te draaien, waarna weer een nieuwe member van het ensemble gereed was. Deze manier van ensemble voorspellen wordt vaak Ensemble Prediction System genoemd. Een andere manier van ensemble voorspellen is om voorspellingen van verschillende beschikbare meteorologische of hydrologische modellen en aanbieders te verzamelen en naast elkaar te zetten. Dit type wordt Poor-Man's of Multi-Model ensemble genoemd. Een derde Ensemble Prediction methode is het random samplen van parameterwaarden en invoergegevens binnen de onzekerheidsband en het model vervolgens opnieuw te draaien. Deze laatste techniek is tot dusver de meest toegepaste methode bij hydrologische ensemble modellering.

Het in deze dissertatie ontwikkelde raamwerk voor het toepassen van meerdere gegevensbronnen in hydrologische modellering is gebaseerd op een manier van ensemble simulatie die verwant is aan het Multi-Model ensemble. In dit raamwerk wordt elke beschikbare bron gebruikt om stroomgebiedsparameters of invoerreeksen af te leiden. Elke unieke combinatie van parameterwaarden en invoergegevens leidt vervolgens tot een andere hydrologische modelsimulatie: een nieuw lid van het ensemble (Ensemble Member). De members vormen samen een ensemble van hydrologische simulaties. Door deze benadering te volgen, worden alle beschikbare bronnen op een effectieve manier gebruikt: er gaat geen informatie verloren. Het raamwerk resulteert in een hydrologische ensemble simulatie die de doorwerking van een deel van de onzekerheid in modelinvoer- en parameterwaarden kwantificeert. Door een relatief gewicht toe te kennen aan iedere member kan de kansverdeling van de ensemble simulatie worden benaderd, alsook de verschillende momenten van die distributie waaronder het ensemble gemiddelde.

Naast het effectieve gebruik van meerdere gegevensbronnen voorziet het raamwerk ook in het toepassen van verschillende technieken voor Data-Model Integration, zoals het invullen van hiaten, datacorrectie, data-merging, data-assimilatie en het aanpassen van data op basis van modelresultaten (feedback loop). Elke alternatieve techniek voor data-model integration leidt weer tot een unieke hydrologische simulatie en dus weer tot een extra ensemble member. In dit onderzoek zijn de volgende data-model integration technieken toegepast: standaard gebruik als modelinvoer, merging meteorologische data alvorens als modelinvoer te gebruiken, het opvullen van hiaten in Earth Observation data op basis van modelresultaten en data-assimilatie.

Om structurele of dynamische (per seizoen of in natter wordende of verdrogende omstandigheden) verschillen in nauwkeurigheid van de ensemble members in acht te nemen, kunnen verschillende methodes worden toegepast voor het toekennen van de gewichten (bijv. Static Weighting, Dynamic Weighting, Model Committees). De methodes voor het toekennen van gewichten kunnen leiden tot verbeterde kansverdelingen van de hydrologische simulatieresultaten. In dit onderzoek zijn Static Weighting en Dynamic Weighting toegepast.

Het uiteindelijke resultaat van het Data-Model Integration raamwerk - na toepassing van meerdere gegevensbronnen, integratietechnieken, en methodes voor toekennen van gewichten - is een hydrologische ensemble simulatie.

Het raamwerk is getest met een gedistribueerd hydrologisch model van het beheergebied van Rijnland in Nederland. Het model is ontwikkeld in SIMGRO. SIMGRO is geschikt voor het modelleren van watersystemen met polders en regulerende kunstwerken (bijv. stuwen en pompstations) zoals dat van Rijnland. Er zijn 24 ensemble members gemaakt op basis van drie landgebruikskaarten: LGN6 als lokaal product, LANDSAT 5 en GlobCover gebaseerd op satellietgegevens, twee bodemkaarten: de Nederlandse en een Europese data set, twee bronnen voor neerslagmetingen: grondstations en weerradar, en twee gegevensbronnen voor verdamping: op basis van weerstations en satellietgegevens van Terra/MODIS. Het resulterende ensemble van afvoersimulaties, de individuele members en het ensemble gemiddelde, zijn vergeleken met de gemeten afvoer van oktober 2012 t/m oktober 2013. De simulaties zijn kwalitatief en kwantitatief geanalyseerd op basis van afvoergrafieken en deterministische en probabilistische indicatoren. Van de deterministische indicatoren zijn de Nash-Sutcliffe Efficiency (NSE), Percent Bias (PBIAS) en de correlatiecoëfficiënt berekend. Van de probabilistische indicatoren zijn de Brier Score (BS) en Relative Operating Characteristics (ROC) bepaald.

De afvoergrafieken voor het praktijkonderzoek van Rijnland laten zien dat de nauwkeurigheid van de individuele members varieert met de tijd, waardoor er geen member kan worden aangewezen die consistent beter is dan de andere members. Een gevoeligheidsanalyse laat zien dat onzekerheid in parameterwaarden als gevolg van de verschillende gegevensbronnen voor stroomgebiedskarakteristieken, minder invloed heeft op de scores van de resulterende afvoersimulaties dan de onzekerheid in modelinvoer op basis van de verschillende gebruikte bronnen voor neerslag en verdamping.

De kwantitatieve analyse van de afvoersimulaties laat zien dat het ensemble gemiddelde beter presteert (met een NSE van 0.85 en PBIAS van 13.1%) dan de meeste individuele members, met inbegrip van het basismodel (met een NSE van 0.81 en PBIAS van 22.2%). Ondanks dat sommige individuele members voor een bepaalde indicator beter presteerden over een bepaalde periode, levert het gemiddelde van het ensemble over het algemeen een nauwkeurigere schatting van afvoer op.

Het gebruik van het complete ensemble als probabilistische simulatie (door een gelijke kans, gewicht, toe te kennen aan elke member) levert goede resultaten op voor het aangeven van overschrijdingen van afvoergrenswaarden in Rijnland. Hoge ROC-scores (bijv. 0.80 voor het 90ste afvoerpercentiel) laten zien dat overschrijdingen van afvoergrenswaarden aangegeven kunnen worden met hoge Hit Rate (succesfrequentie) en lage False Alarm Rate (frequentie van het aangeven van een overschrijding terwijl die in werkelijkheid niet optreedt). De ROC-scores duiden op een voordeel van het gebruik van de kansverdeling van de ensemble simulatie ten opzichte van het gebruik van uitkomsten van individuele simulaties. Dit laat voor het praktijkonderzoek van Rijnland, ondanks het beperkte ensemble van 24 members, zien dat met de hier gepresenteerde benadering probabilistische simulaties kunnen worden gegenereerd die effectiever zijn in het aangeven van afvoeroverschrijdingen dan deterministische benaderingen waarbij slechts een deel van de beschikbare gegevensbronnen wordt gebruikt.

Data Assimilation (DA) is toegepast als methode voor data-model integratie voor actuele verdamping van Terra/MODIS. DA met Particle Filter en Residual Sampling leidde tot een verbeterde afvoersimulatie voor Rijnland ten opzichte van het basismodel.

Er zijn twee methodes getest om gewichten toe te kennen aan de ensemble members met als doel het gemiddelde van de ensemble afvoersimulaties verder te verbeteren. Static Weighting op basis van nauwkeurigheidsindicatoren van de individuele members leidde niet tot een verbetering. Dynamic Weighting, waarbij gewichten verschillen met de tijd op basis van dynamische verschillen in nauwkeurigheid van de members en waarbij, ook dynamisch, geen gewicht wordt toegekend aan de minst presterende members, leidde wel tot een verbetering van het gemiddelde van het ensemble ten opzichte van het toepassen van hetzelfde gewicht aan elke member.

Op basis van de hierboven gepresenteerde resultaten voor het praktijkonderzoek van Rijnland wordt geconcludeerd dat het ontwikkelde raamwerk voor het gebruik van meerdere gegevensbronnen in hydrologische modellering, geïnspireerd op de Multi-Model Ensemble benadering; effectief kan worden toegepast, kan leiden tot verbeterde afvoersimulaties en de effecten van een deel van de onzekerheden in modelparameters en invoergegevens kan laten zien. Wanneer deze aanpak wordt gecombineerd met de veelgebruikte ensemble technieken van paramater sampling, modelleren met verschillende modelstructuren, en invoer van meteorologische ensemble voorspellingen, kan een volgende stap worden gezet om waterbeheerders te voorzien van betrouwbare hydrologische ensemble simulaties en voorspellingen.

About the author

Isnaeni Murdi Hartanto was born in Kendal, Central Java Province, Indonesia. He likes to read and playing PC games in his spare time, and has great curiosity in computers, physics and astronomy.

He finished his bachelor in civil engineering at the Engineering Faculty of Diponegoro University in 2001. In 2008, he started his MSc at UNESCO-IHE, Delft, the Netherlands, and graduated in 2010 with distinction. His MSc thesis was about using coastal hydrodynamics software on a fluvial environment, showing that a specially build software can be applied to a different environment, of course with caution. This work was published in a peer reviewed journal in 2011.

In 2011 he started his PhD research in Hydroinformatics at UNESCO-IHE, on integrating different data sources in hydrological models. The research was part of the MyWater Project from EU FP7 programme, which aimed to build an information platform for water managers by delivering hydrological data from earth observation and hydrological models in a comprehensive web client. During his research, he worked with companies and people across the globe involved in the project. The international connections excited him, he had the opportunity to work with Dutch, Greek, Portuguese and Brazilian colleagues. His research was mainly dealing with data processing, hydrological models, satellite products, software, coding, and field surveys, all things that he likes. During his PhD research, he also supervised two MSc students in writing their master thesis.

List of Publications

International peer-reviewed papers

- I.M. Hartanto, L. Beevers, I. Popescu, N.G. Wright (2011) Application of a coastal modelling code in fluvial environments. Environmental Modelling & Software 26:1685–1695
- I.M. Hartanto, C. Almeida, T.K. Alexandridis, M. Weynants, G. Timoteo, P. Chambel-Leitão, A.M.S. Araujo (2015) Merging earth observation data, weather predictions, in-situ measurements and hydrological models for water information services. Environmental Engineering and Management Journal 14(9)
- T.K. Alexandridis, I. Cherif, G. Bilas, W. Almeida, I.M. Hartanto, S.J. van Andel, & A. Araujo (2016). Spatial and Temporal Distribution of Soil Moisture at the Catchment Scale Using Remotely-Sensed Energy Fluxes. Water, 8(2), 32
- I.M. Hartanto, J. van der Kwast, T. K. Alexandridis, W. Almeida, Y. Song, S.J. van Andel, & D.P. Solomatine (2017). Data assimilation of satellite-based actual evapotranspiration in a distributed hydrological model of a controlled water system. International Journal of Applied Earth Observation and Geoinformation, 57, 123–135.
- I.M. Hartanto, I. Cherif, T.K. Alexandridis, J.J. Reitsma, S.J. van Andel, D.P. Solomatine, Ensemble simulation based on multiple data sources in distributed hydrological modelling (submitted)

Conferences papers

- I.M Hartanto, S.J. van Andel, D.P. Solomatine, Ensemble simulation from multiple data sources in spatially distributed hydrological model of the Rijnland water system in The Netherlands, *Hydroinformatics Conference 2014*, 17-21 August 2014, New York City, USA 2014

- N. Silleos, S. Strati, I. Cherif, C. Topaloglou, T.K. Alexandridis, C. Iordanidis, D. Stavridou, S. Monachou, C. Kalogeropoulos, G. Bilas, N. Misopolinos, T.F. Chiconela, W.G de. Almeida, I.M. Hartanto, S.J. van Andel, A.A. Nunes,. and P.C. Leitao: Weekly time series of LAI maps at river basin scale using MODIS satellite data, *in Proceedings of 1st International GEOMAPPLICA Conference*, 8-10 September 2014, Skiathos Island, Greece, Skiathos Island, Greece., 2014

Selected other publications

- I.M Hartanto, S.J. van Andel, A. Lobbrecht, A. van Griensven, D.P. Solomatine, Integrating Earth Observation Data Into Hydrological Modeling And Water Management, 22-27 April 2012, EGU general assembly 2012
- I.M Hartanto, S.J. van Andel, A. Jonoski, D.P. Solomatine, Hydrological modelling of low-lying catchments in deltas using multiple data sources and SIMGRO modelling system, 7-12 April 2013, EGU general assembly 2013
- S.J. van Andel, I.M. Hartanto, Merging Earth Observation Data, Weather Predictions, In-situ Measurements, And Hydrological Models, For Reliable Information On Water, 30 Oct - 1 Nov 2013, 14th WaterNet/WARFSA/GWP-SA Symposium
- I.M Hartanto, S.J. van Andel, A. Jonoski, D.P. Solomatine, Spatially distributed modeling of controlled low-lying catchments utilizing earth observation data, 9-13 December 2013, AGU Fall meeting 2013
- H. van der Kwast, I.M. Hartanto, S.J. van Andel, 2013: Integrating in-situ measurements and remote sensing data in process-based hydrological modelling, VUB, Brussels, Belgium

- I.M Hartanto, S.J. van Andel, T.K. Alexandridis, D.P. Solomatine, Feedback Loop of Data Infilling Using Model Result of Actual Evapotranspiration from Satellites and Hydrological Model, 27 April-2 May 2014, EGU general assembly 2014
- Y. Song , I.M Hartanto, S.J. van Andel, T.K. Alexandridis, D.P. Solomatine, Frequency analysis of Earth Observation and hydrological model estimations of evapotranspiration and soil moisture, 27 April-2 May 2014, EGU general assembly 2014
- I.M Hartanto, S.J. van Andel,, T.K. Alexandridis, A. Jonoski , D.P. Solomatine, Validation of multi-input ensemble simulation with a spatially distributed hydrological model in Rijnland, the Netherlands, 27 April-2 May 2014, EGU general assembly 2014
- T.K. Alexandridis, A. Araujo, P. Chambel Leitao, I. Cherif, D. Stavridou, S. Strati, C. Iordanidis, A. Nunes, I. Hartanto, W. Almeida, S.J. van Andel, G. Bilas, N. Silleos and N. Misopolinos, Estimating Weekly Time Series of Hydrological Information (LAI, ETa and soil moisture) using Satellite Remote Sensing, , 20th-23rd October 2015, Earth Observation for Water Cycle Science 2015, ESA-ESRIN, Italy